# Why We Do It

ALSO BY NILES ELDREDGE

*Reinventing Darwin: The Great Debate at
the High Table of Evolutionary Theory*

*Life on Earth: An Encyclopedia of Biodiversity,
Ecology and Evolution*

*The Triumph of Evolution:
And the Failure of Creationism*

*The Pattern of Evolution*

# Why We Do It

*Rethinking Sex*

*and*

*the Selfish Gene*

NILES ELDREDGE

W. W. Norton & Company
NEW YORK • LONDON

Copyright © 2004 by Niles Eldredge

All rights reserved
Printed in the United States of America
First Edition

For information about permission to reproduce selections from this book,
write to Permissions, W. W. Norton & Company, Inc., 500 Fifth Avenue,
New York, NY 10110

Manufacturing by Courier Westford
Book design by Brooke Koven
Production manager: Julia Druskin

Library of Congress Cataloging-in-Publication Data

Eldredge, Niles.
Why we do it : rethinking sex and the selfish gene / Niles Eldredge.—1st ed.
p. cm.
Includes bibliographical references.
**ISBN 0-393-05082-3 (hardcover)**
1. Sex (Biology) 2. Sociobiology. 3. Human evolution. I. Title.
QP251.E42 2004
155.3—dc22

2003027564

W. W. Norton & Company, Inc.
500 Fifth Avenue, New York, N.Y. 10110
www.wwnorton.com

W. W. Norton & Company Ltd.
Castle House, 75/76 Wells Street, London W1T 3QT

1 2 3 4 5 6 7 8 9 0

*In the spirit of Marvin Harris, 1927–2002*

# Contents

*Part One*

# The
# Duality
# of Life

# ONE

# Obsessed
# with Genes

Why do people have sex? A no-brainer, you might say: people have sex because it feels good. It's fun. And temporary satiation, as with any appetite, is soon overcome with the urge to do it again.

But hold on, surely there's more to it than that. There are *consequences* of having sex: there's the risk of disease, there's the risk of falling in love (hence the risk of "commitment"), and there is, of course, the risk of pregnancy. No one, presumably, has sex in order to contract a social disease, but there are plenty of anthropologists and psychologists who have argued that having sex forges a "social contract." And there are many others, including leaders of the Catholic Church and evolutionary psychologists, who insist that the primary, even the sole, purpose of having sex is to make babies.[1]

Bill Clinton claimed that his interaction with Ms. Lewinsky

did not fit his definition of "having sex," evidently reserving that term strictly for sexual intercourse. But when church fathers and evolutionary psychologists insist that the primary reason people have sex is to make babies, they unconsciously raise a host of related issues: Why, for example, does homosexuality exist? What about masturbation, fellatio, cunnilingus, and other forms of "sodomy"? What is rape all about? For these are thoroughly sexual acts, yet have for the most part absolutely nothing to do with making babies. And though the answer to at least some of these questions may in part be "to avoid the consequences" of unprotected copulation, just think of the consequences for Clinton, Lewinsky, Congress, and the American people of their sexual encounters.

So maybe "why people have sex" is not as straightforward a proposition as it first seems. Maybe the reasons why people have sex, and engage in all manner of sexual behavior, is more complicated than simple baby making or than just having fun. We know that sex is an almost universal characteristic of the living world: even most bacteria, which usually reproduce by simply splitting into two clonal copies of their "parent," nonetheless also occasionally exchange genetic information, perhaps the most elemental definition of sex imaginable. And humans, as mammals, are no different in this regard from barnyard livestock. Since time immemorial, humanity has had no trouble realizing that, however much some of us might insist on having been created in God's image, down deep we are undeniably animals—for we hump just like the pigs in the wallow.

Or so it seems. I well remember my ninth-grade sex education film, which at one point (much to our delight) homed in on an enormous pair of pig's testicles to make the point that, anatomically speaking, all the parts of pigs, horses, cows, and humans seem to match up pretty well. So far so good; and it does seem to be an incontrovertible fact that, when female lions in the wild, or pigs in the barnyard, go into "heat," one or more males soon appear to mount her. But it is only then, in estrus, that female mammals both of the barnyard and in the wild are receptive. No heat, no sex. Estrus triggers sex and the straightforward "goal," to everyone's evident satisfaction, is to make more lions or more pigs. Yet already in ninth grade, the girls (and even some of the boys) seemed to realize that their nascent interest in the opposite sex was not based on the time of the month.

## The Myth of the Self-Spreading Gene

Ever since the "molecular revolution" triggered by Watson and Crick's unraveling of the structure of DNA, genes have been the dominant metaphor underlying explanations of all manner of human behavior, from the most basic and animalistic, like sex, up to and including such esoterica as the practice of religion, the enjoyment of music, and the codification of laws and moral strictures. Read virtually any issue of the Science Times (published by the *New York Times* on Tuesdays), and you are likely to find an article announcing the "discovery" that homosexuality, or intelligence, is genetically based—or that it is genes, after all, that supposedly make men better than women at math. The old nature/nurture debate, in other words, has been tilted heavily toward the nature side when it comes to the human condition—an overreaction, as we'll see, to recent progress in deciphering the genetic code.

The media are besotted with genes. You can't watch a nature show on television without hearing the narrator solemnly telling you that the snake slithers into the bird's nest, or the shark sleeplessly prowls the sea, each on the hunt—but that

their real, underlying motive is to get enough energy to reproduce. According to the popular press and television, *everything*,
every bit of animal behavior, really boils down to passing genes
along to the next generation. Forgotten in the process is the
simple fact that animals need to eat simply to live.

Consider the reinterpretation of animal and plant domestication that has crept into the popular press in recent years. For
at least the past fifteen thousand years, humans have been pick-
ing and choosing which of their dogs, cattle, sheep, and stands
of wheat they will allow to reproduce—in the reasonable expec-
tation that the features (coat length, milk yields, etc.) they
admire most in those chosen to reproduce will be present to an
even greater extent in the offspring. Darwin knew about selective breeding, studied it, and used his understanding to help for-
mulate the notion of natural selection—cornerstone of his
theory of evolution. But now some writers, inspired by the pre-
sumption that there is actually a competitive race to leave
copies of genes to the next generation, and that this race is what
the game of life is all about, have begun to suggest that domestication is really a clever act—not of humans, but of domesticated plants and animals themselves!

For example, the *Newsweek* writer Stephen Budiansky wrote
an entire book (*The Covenant of the Wild: Why Animals Chose
Domestication*, 1992) saying, in effect, that domesticated animals deliberately brought themselves in out of the Ice Age cold,
thus avoiding the fate of extinction that befell so many of the
adamantly wild larger mammals, such as wooly rhinos and mammoths, mastodons, giant bison, and many other species. More
recently, a *New York Times* Op-Ed piece[2] sees domestication as
a mutually advantageous "dance" of species. Focusing on corn
(*Zea mays*), the essay tells us that the corn gets to have its genes
moved all around the world, in exchange for the "gratification

of human needs," and at the expense of habitats where other species might thrive and at the added cost of the pollution that comes from modern agricultural practice. As if a corn plant cared about spreading its genes instead of just doing what it was told (with genes, increasingly, that are artificially added from other sources). Nor is this rhetoric particularly tongue in cheek.

So there's an assumption, now widespread and tacitly accepted in our culture, that there is an ineluctable urge, a necessary force, that causes organisms to seek to spread their genes—or genes to spread as many copies of themselves as possible. And there is an even deeper assumption that this quest for genic dispersion amounts to life's "bottom line"—the ultimate reason why all things from bacteria to humans do what they do. We need to look at these assumptions very carefully, ask where they come from and how accurate a picture of life they present: for though I am a parent of two children whom I love very much, I do not think for one moment that everything I have ever said or done in my life has been geared—directly or even indirectly—toward having kids.

## ORIGIN OF A MYTHIC OBSESSION: GENES IN ACADEME

We can't pin the blame for this pervasive cultural obsession with genes on the news media and entertainment industries. They are just giving us an oversimplified picture of what has become the dominant line of thinking in modern biology. Especially in evolutionary theory, genes have for over half a century easily eclipsed the outside natural world as the primary driving force of evolution in the minds of many evolutionary biologists. This is the nature/nurture debate raised up one level: instead of focusing on how much of human behavior is learned (nurture) versus how much is dictated by genes (nature), the debate in evolutionary biology centers on the relative importance of the urge to spread genes (in this case, nature) versus the role played by the external environment (nurture) in promoting stability and change in the history of life. Sure, the environment might *seem* to be important, but in reality, most evolutionary biologists say, environmental factors are simply direct stimuli; the real, "ultimate" cause of evolutionary change comes from the built-in, incessant urge of genes to beat out other genes, to make it to the next generation. One geneticist has even said, genes "don't pay attention to the weather."

So, everything we see—whether it is wildebeests migrating over the African savannas, owls taking voles at night, elks rutting in the moonlight, or Beatles singing on Ed Sullivan's show—is routinely reduced to this primary purpose of genes doing their best to push one another aside in the race to reach the next generation. E. O. Wilson's notion of "consilience," what he sees as a great synthesis (literally a "jumping together") of all the hallowed disciplines of human thought in the humanities, social sciences, and natural sciences to form one integrated approach to explaining life and especially the human condition, is in reality his move to explain virtually everything as the outcome of genetic competition and genetic expression. Evolutionary biology, it seems, is out to eat everyone else's lunch, as the role played by the humanities and social sciences in the explanation of who humans are and how they came to be takes a backseat to evolutionary biology—specifically, its narrowest, gene-centered form.

This rapture over the power and importance of genes in both daily life and evolution has actually come from just one corner of evolutionary biology. As molecular biology—the study of the chemical structure and functions of DNA and RNA, genes and chromosomes—began to take hold and command the attention of more and more biologists (and funding agencies), evolutionary genetics found itself in a tight corner. The new molecular genetics was stealing all the thunder from the older "population" genetics (the mathematical analysis of the fate of genes in populations) of traditional evolutionary theory. Worse, it looked as if it would be the new breed of geneticist who would have all the fun *and* all the money—explaining in the dazzling new language of molecules not only why organisms look like their parents but probably also that new knowledge translates into an explanation of how the evolutionary process actually works.

Evolutionary geneticists, naturally enough, reacted to this perceived threat from molecular biology. Molecular biology is expensive. Partly because of molecular biology's unslakable thirst for space and expensive equipment, many old biology programs split into two or three separate departments, including the by now nearly ubiquitous "Department of Ecology and Evolution." Many of these new programs devoted to ecology and evolution themselves became, at least in good measure, "molecularized." And this was a very good thing indeed—since genes, of course, do lie close to the heart of the evolutionary process—and have much to tell us, as well, about the course of evolutionary history.[3]

Yet evolutionary biology's reaction to the competition—over money, space, and glamour—sparked by the new molecular biology did not stop with declarations of political independence and the building of new campus buildings. Unsurprisingly, and more interestingly, the riposte to the molecular world offered by the challenged world of population genetics also took the form of revised theory. Population geneticists, the inheritors of a rich intellectual tradition going back to the early twentieth century (peopled by such storied greats as R. A. Fisher, J. B. S. Haldane,

Sewall Wright, and Theodosius Dobzhansky), understandably decided to head off molecular biology's inroads and threats at the pass. In effect, they found ways of, if not eating molecular biology's lunch, at least assuming leadership in interpreting how molecular biology could be incorporated into the mainline version of evolutionary theory of the late 1950s.

Already, molecular biologists were asking the inevitable question: What does this exciting new field of research on the structure and function of DNA and RNA tell us about how evolution works? By 1972, the molecular biologist[4] Jacques Monod had published his creative and challenging reflections on the evolutionary ramifications of molecular biology in his *Chance and Necessity*. Many working molecular biologists began following suit, some of them positing novel forms of evolutionary mechanisms with unfamiliar names such as "molecular drive."[5]

So, beginning in the 1960s, evolutionary biologists began to revamp old-style Darwinian evolutionary notions—especially "natural selection"—expressly in terms of genes. This is where the excesses began. Most famous of these gambits is surely the notion of "selfish genes," named and popularized by the British biologist Richard Dawkins in the mid-1970s. Dawkins argues that genes actually vie with one another for representation in the next generation, and that the more successful ones will be present in proportionally larger numbers in the next generation. Natural selection, to biologists like Dawkins, boils down to a struggle for organisms to reproduce, to send their genes along to the next generation. And evolution is simply a fallout of this process: as the ages roll, and generations come and go, the immortal genes are passed along, with the superior versions of them always winning out and changing the complexion of life as time goes by.

## PHYSICS ENVY

In the next chapter, I'll have much more to say about the scientific content and validity of this emphasis on genetic competition as the latest interpretation of natural selection (and thus, in many quarters, nearly synonymous with "evolution"). But there were broader issues concerning what science is, and how it is to be done, that were also in play as molecular biology posed its territorial threat to traditional evolutionary genetics.

Science is the study of the material contents of the universe and the interactions among them—to paraphrase a possibly apocryphal definition often attributed to the physicist Ernst Mach. In that light, evolutionary biology is a perfectly good science, posing and testing causal theories (like natural selection) against experimental and historical data. On the other hand, evolution has never seemed to be quite like the archetypical science of physics. Darwin himself was dismayed—and rightly so—when the prestigious English physicist John Herschel pronounced natural selection the "Law of Higgledy-Piggledy." And later evolutionary biologists, such as Ernst Mayr, have occasionally wondered aloud why physicists can't seem to grasp the concept of natural selection. Inasmuch as physicists are not a uniformly stupid lot, that many of them don't seem to "get" core evolutionary concepts like natural selection is actually very interesting.

One possible reason for this persistent misunderstanding may lie in the very nature of the original, traditional formulation of natural selection. It is very un-physics-like: it says that how well an organism fares relative to its peers in eking out an existence in a world of finite resources has, in all likelihood, an effect on its ability to reproduce. Those elephants best able to find and exploit young palm trees, grasses, and other delectables on the elephant menu will tend to thrive more than their less successful counterparts and, other things being equal, will tend to have more sex and be more successful at making babies.

True, physicists have long since grown accustomed to thinking statistically (witness the gas laws, radioactive decay rates, quantum mechanics); but natural selection as originally conceived is also a fundamentally *historical* process: it is the process of recording, on a ledger book now known to consist of genes sitting on chromosomes, what works better than what as organisms compete for food resources, territories, and mates. This sort of process is very different from that of the Moon's gravitational pull on Earth, or the interactions of elements to form chemical compounds. Somehow, the release and capture of energy in physical systems seems so *active*. In contrast, natural selection seems so disturbingly *passive*. So un-science-like.

There is matter and energy galore flowing through biological systems. But it is in the bodies of organisms and their interactions with other organisms and the physical world, in the context of ecosystems, where all that matter and energy flows. Genes, in contrast, are about storage and utilization of information. And though "information theory" was conceived by the likes of John von Neumann and Claude Shannon over half a century ago, using mathematical constructs very much like those of thermodynamics, this line of thought has yet to pervade biological thinking—despite some interesting early attempts. A true physics of genetic information remains to be developed.

But there *is* the selfish gene. The selfish gene is imagined to be an active player in the game of life: no mere bit of stored information, but a force to be reckoned with. It is pictured as an active particle (or at least the bit of information stored on that particle) seeking to transcend its role as something need-ing translation by another molecule (RNA) to build, for exam-ple, a hemoglobin molecule in an animal's body. The selfish gene is out for bigger game: to see to it that as many copies of itself are disseminated into the germ lines of descendant organ-isms, there to be disseminated still farther. The selfish genes are the imperialists of the biological world.

True, in physics there are no known molecules, atoms, or subatomic particles thought to behave for their own benefit, as if they had motives, desires—even minds—of their own. But consider this: in molecular biology, genes do things; they copy themselves, and they make themselves available to be copied by RNA in the ongoing process of making enzymes and other gene products. Molecular biology seems so very *scientific*, not just because it uses fancy laboratory equipment and proce-dures—and eats up so much money. Molecular biology actually studies the way things work, how bits and pieces of nature inter-act to form other bits and pieces of nature.

Hence the selfish gene: it seems to *do* something, play an active role in deliberately trying to maximize its spread, in a way at least a little like the way genes make products inside test tubes and organisms' bodies. The selfish gene seems to be the evolutionary counterpart to the gene of molecular biology—and, not coincidentally, to present a more conventionally "sci-entific" look to evolutionary biology. How well it actually helps us understand the evolutionary process is another matter, taken up in some detail in the ensuing chapters.

## BEYOND THE SELFISH GENE

At about the same time (the mid-1970s) when Dawkins wrote his *The Selfish Gene*, the Harvard biologist E. O. Wilson coined the term "sociobiology" for his wedding of selfish-gene-style thinking to the nature and evolution of animal social systems. Darwin was one of the first to see a contradiction between the selfish competitive struggle between organisms in the "struggle for existence" (as he put it) and the cooperative behavior exhibited by social creatures such as ants, monkeys, . . . and humans. The British biologist William Hamilton, in two important papers published in the 1960s, had established that organisms could be expected to cooperate with one another to the degree to which they share their genes. Wilson and others took this resolution of Darwin's paradox and ran hard with it— essentially founding an entire school of research based on the premise that social systems are breeding cooperatives all fundamentally concerned with the perpetuation of the underlying genes that dictate the structure of the society and the behaviors of the component organisms.

Wilson's book *Sociobiology* (1975) extended the argument to human behavior. It touched off a storm. No one except a subset of academics cared about Wilson's application of selfish-gene-style thinking to animal social systems. But human behavior? That was different! Social scientists were offended that their careful dissections of the fabric of human life were ignored by this sweeping new explanation of everything human in terms of a deep-seated desire to spread our genes around—no different from the controls of behavior of all other organisms on Earth. Some anthropologists became converts, actively adopting sociobiology as their ruling paradigm in interpreting the vast spectrum of human ways of being.

Over the years, the human side of sociobiology has transformed into "evolutionary psychology." And it is interesting, as we'll see in later sections of this narrative, that evolutionary psychologists, couching their approach to human behavior in explicitly selfish-gene-based evolutionary biological terms, repeatedly claim that theirs is the more "scientific" approach to understanding the root causes of human behavior.

I have called this gene-centered approach to evolutionary thinking "ultra-Darwinism." Its broader cultural influence is not to be taken lightly. I have seen recent reports in the media, for example, that there is a new school of historians—"bio-historians"—who are determined to reinterpret history as the outcome of actions of individuals whose "fitness" (success at spreading their genes) was enhanced by their actions. Don Juan, possibly. Perhaps even Attila the Hun. But Gandhi?

## PLAN OF ACTION

Selfish genes, sociobiology, evolutionary psychology, the linkage of all three with the revolution in molecular biology, and leakage of all of this into the news media and popular culture—all this contributes to our tendency to think that people have sex for the most part to make babies. We have come to be besotted by genes, seeing them as the masterminds of the biological universe and as the determinants of all that is human. But I have a very different view of the very nature of life, of evolution, of the nature of social systems, and what it is to be human. Above all, why it is that people have sex.

Much of this gene-infested biology that has led to a gene-infatuated culture is simply a distorted description of the structure and function of the living world—of the biological components of the contents of the universe and the interactions between them. My aim in this narrative is to provide an alternative—a description of the natural world and how it works that is better balanced and more accurate than Dawkins's vision. I will use plain speak to do so, confident that what I have to say is basically simple, and so can be expressed in simple and direct language.

- I will restore the economic side of life—making a living—to its rightful place as at least an equal to sex (in the sense of baby making and gene spreading) both in the lives of organisms and, more generally, in the evolutionary process.

- I will tell you that organisms—all organisms, including human beings—have an economic and a reproductive side to their lives. Every living creature is part of at least two larger-scale systems that are hierarchically structured.

- I will tell you that Darwin's original description of natural selection—as the effect that success in an organism's economic life has on its reproductive life—remains far and away the best way of thinking about this, the central evolutionary process, better than Dawkins's "selfish gene" version, which, among its other problems, excludes explicit reference to the environment.

- I will also show you that genes don't drive evolution. Far from it: nothing much happens in evolution unless environmental change—real, substantial environmental change—shakes up ecosystems and drives species extinct. Only then does anything new happen under the evolutionary sun.

- I will then say that the flawed gene-imbued version of evolution works in social systems only to the extent that social systems are composed of groups of closely related organisms. It works best for colonial marine organisms like corals, and for the bees and ants for which it was originally invented. It is a dicey proposition, though, to apply such terms with unalloyed success to vertebrates. And humans, who form a myriad of different (and hierarchically structured) groups simultaneously, fit the profile least.

- Sex is so clearly separated from pure reproduction in humans—and there is so much interplay between sex and economics, and even between economics and reproduction in human life—that this "human triangle" of sex, reproduction, and economics makes us the very least likely creatures on the planet to conform to such strictures of evolutionary determinism.

- Traditional humanism and social science have been more right than wrong all along. Culture has increasingly guided our way to economic (and reproductive) behavior over the past 2.5 million years. Culture frequently runs counter to, and overrides, biological drives and processes. Living outside of local ecosystems—the bailiwick of natural selection—we are so atypical, so fundamentally *not* subject to the biological rules that still guide the lives of all other species that to reduce our existence in this manner, to see ourselves as mere shells being marched around by our inner genes, is not just bad biology. It verges on being a willfully stupid joke or, even worse, a malevolent political doctrine.

- I will also say that this fundamentally flawed biology, which has increasingly swept over both academic and general cultural thinking about who we are and why we do the things we do, has profoundly bad implications for social theory and its political implementation. Eugenics may have been a naïve, and even idealistic, movement when it was founded by Darwin's nephew Francis Galton. But many people were sterilized in its name, and its underlying thinking shows up in *Mein Kampf*—an indication that gene-based biology is similarly flawed and that the social policies that we adopt should never be based on supposed inevitable truths about genetics, or supposed genetic differences between ethnic groups or social strata.

So here's my plan: I will start by looking at how life is organized. What do organisms do? Why? Why, especially, do organisms reproduce? Why is reproduction among complex organisms like animals and plants almost universally sexual? What is natural selection all about, anyway? And how does the evolutionary process really work? What are social systems? How do they work? How do they evolve? And then the big question: What does all this mean for understanding human life? And why, after all, do people really have sex?

In effect, I will be retracing the footsteps along the path that led to selfish genes → sociobiology → evolutionary psychology, rethinking each and every step of the way—correcting all of it as I develop a more neutral, less value-laden description of what life is all about.

# TWO

# Chickens and Eggs

## The Two Sides of Life

To understand why people have sex, we need to dig a bit deeper. We'll need to ask some basic questions: What do animals—and plants, fungi, and microorganisms—actually *do*? If they have sex to make babies, what are some of the consequences of baby making? What else do animals do? Why do animals eat, breathe, urinate, and defecate? And what are the consequences of these most elemental of acts? Gene-centered evolutionary biologists will tell you that animals eat and do all those other things so that they can have sex, and they have sex to make babies, to pass along their genes, because it is the genes themselves that are selfishly trying to ensure their presence in the next generation. But what if we just look at what organisms actually *do*, and see how these activities fit in with one another, without assuming in the first place that one set of behaviors is either *for* or more important than the other?

According to "selfish gene" evolutionary biologists, animals live to reproduce. A hen is an egg's way of making another egg. And it's supposedly no different for humans. But to me it seems as sensible to say that an egg is a hen's way of making another hen. I'm a traditionalist, still scratching my head over the chicken-and-egg dilemma—now transformed from a strictly historical question (which came first?) to a functional question of how the system works. Sounds trivial, perhaps even silly, but that old dilemma completely captures the essence of the problem. The ultra-Darwinian, selfish-gene position essentially says that the instructions for building a system (the genes) are more important than the system itself (the organism); the system exists only, after all, because there is a blueprint for it (genes to make organisms, e.g.)—and the only "purpose" of the system is to ensure that those instructions are passed along.

## The "Meaning" of Life

Science, of course, doesn't deal with strictly philosophical questions, and even philosophers seldom wrangle anymore over life's meaning or purpose. Meaning-of-life issues have long since been relegated to the provinces of received doctrine in organized religion, the eponymous Monty Python movie, and college dormitory bull sessions. But thinking that life is fundamentally all about the spread of genes from one generation to the next comes perilously close to attributing a "purpose"—thus a "meaning"—to all living systems, even though that drive to spread those genes is itself (correctly) seen as essentially mindless.

At stake are two competing underlying assumptions of the fundamental nature of living systems. Scientists do not like to acknowledge that assumptions underlie even the most routinely banal aspects of their daily work. And they *really* don't like to

be told that their entire theoretical edifice rests on largely unexamined assumptions.

Yet I do say it here: the proposition that organisms exist fundamentally to propagate the information from which their bodies are built is a largely unexamined assumption. We have already explored (chapter 1) some of what I take to be the sociological and cultural forces within science that led evolutionary biologists to this rather extreme position. We need now to examine the intellectual content of this gene-centered view of life.

Scientific assumptions are assumptions primarily because they involve statements about the natural world that cannot easily be tested and (possibly) refuted by simply gathering up some data. But we *can* analyze them, by "deconstructing" organic systems—starting with organisms themselves—teasing them apart to see what makes them tick: how the bodies of complex plants and animals are constructed, and how they work.

## Life's Dichotomy

So what do organisms actually do? And how are their bodies organized to perform these various functions? We've already seen this partial list of basic life functions: it includes eating (obtaining energy and nutrients from the outside world); breathing ("respiration": to most of us, taking in oxygen; to a biochemist, using that oxygen in our cells to burn food and thus release the energy to power cellular functions); digestion (including "absorption": breaking down of food into chemical forms that can be absorbed through the intestinal linings); circulation (of blood, which shunts absorbed food and oxygen to cells and picks up waste products); excretion (filtration and discharge of those poisonous or unneeded by-products of cellular metabolism, like uric acid); and elimination (removal of nonabsorbed solids from the intestines). And now—for something completely different—we add sex, here meaning "reproduction."

There is a fundamental disconnect—a fundamental dichotomy—between reproduction and all the other basic life processes. Look at it this way: think how long you would live if you didn't (1) eat and drink water, (2) digest and absorb, (3) have your oxygenated and nutrient-laden blood circulate, (4) urinate, and (5) defecate. Answer: it varies, but not very long in any case. Political prisoners, sipping only water, waste away pretty quickly. The brain soon shuts down, taking the heart with it, when oxygen is cut off for only a few minutes. Uremic poisoning also sets in fast if the kidneys stop functioning efficiently—and so on down the line. These and all the other subdivisions of economic labor taken up by the "vital organs" of any complex animal's body are just that: literally vital. If one of them stops working, you're dead pretty quickly.

Okay: now imagine how long you or anyone could live without sex. Answer: until something else kills you. There's a deep asymmetry here. A human body—indeed, every kind of organism's body—simply *must* perform all the other bodily functions constantly and continuously just to stay alive—all, that is, save having sex. And though there are those who insist that life is not worth living without it, the truth is that sex, and even more reproduction, is the one physiological arena that is *not* essential to the very act of living.

How reasonable an assumption can it be that we eat to live, but live to reproduce—that we ultimately eat, not to live, but to love? We cannot have sex without living, yet we can live without having sex. August Weismann, an important transitional figure between Darwin and the birth of modern genetics at the turn of the twentieth century, knew (as, at some level, did the farmers in biblical times)[1] that eggs and sperm unite to make babies. From a single-celled fertilized egg comes, after an interval of growth and development, a full-blown human body, now known to consist of billions of cells of several hundred distinct types (skin cells, liver cells, hair cells, etc.). There's something special about those egg and sperm cells, for under the right circumstances they can make every other kind of cell. Weismann pointed out that the reverse is not true: a bone cell cannot make a sperm or an egg.[2]

So here again we have asymmetry, but this time in reverse. Weismann's insight finally put to rest the notion of "inheritance of acquired characteristics" (a notion usually associated with the early pre-Darwinian evolutionist Jean-Baptiste de Monet, chevalier de Lamarck) and (somewhat grudgingly and erroneously) was accepted by Darwin in later editions of *On the Origin of Species*. Weismann was saying that whatever happened to an organism during its lifetime—things like accidentally losing an arm or developing calluses on the hands—would not, *could* not, affect sperm or egg cells, and so could not be passed along to any offspring. Weismann called everything but the eggs and sperm, and the ovaries and testicles that make them, the "soma" (from the Greek for "body"); somatic cells stand in marked contrast to germ-line cells.

The dichotomy might be the same—between the "body" versus "sex cells" and between functions needed to survive versus sex—but in an important sense the tables have been turned. Sex and reproduction are the only actions not necessary to survival, in that sense playing second fiddle to the literally vital bodily functions. Body cells divide to make more body cells of the same type. Only sex cells, upon fertilization (union of egg and sperm in animals), can go on to differentiate into other types of cells, eventually producing a new individual. What happens to sex cells (e.g., mutations in the genes) during the lifetime of an organism can thus be passed along to the next generation, whereas mutations in skin cells cannot.

So here we have a fundamental distinction: between bodies and germ lines, between having sex and all other kinds of biological functions. The vital functions are necessary just to keep the body going: differentiating from a fertilized egg, growing, and simply staying alive. The other is necessary, not for living now, but for being born in the first place—*and* (perhaps!) to make more babies. Let's call one side economics and the other reproduction, marking the gulf between staying alive through procuring energy and processing it internally, on the one hand, and making babies, on the other.

There is a feedback loop between these two categories: reproduction makes new organisms, but the economic side of life houses the genes and supplies the energy for reproduction to occur. Seemingly, 'twas ever thus. As Weismann first told us, the soma is the functioning energetics machine without which there is no life; but only the germ line can make a new soma. Life has been on the planet for over 3.5 billion years—so it is safe to conclude (all flesh being mortal) that, without reproduction, life would not have lasted nearly so long as it manifestly has.

## In the Beginning, There Was Duality . . .

Indeed, even theories of the origin of life present an interesting take on the chicken-and-egg dilemma. The up-to-date version of Weismann's dictum that the germ line makes the soma, but that what happens to the soma during an organism's life cannot affect the genetic information in the sperm- and egg-producing germ line, comes in the form of the so-called central dogma of molecular biology: DNA (acting through the intermediary RNA) specifies the order of amino acids used to construct protein molecules—the workhorse molecules of the soma. Until Tom Cech and his colleagues discovered "reverse transcriptase" (another Nobel Prize–winning piece of research), it seemed that the central dogma was inviolate: genes dictate protein structure, but proteins have no effect on genetic structure. Cech was able to show that the enzyme reverse transcriptase is able to catalyze the synthesis of DNA from an RNA template—raising the possibility that it may have been proteins that came first in the early stages of the origin of life. After all, ever since the famous experiments that Stanley Miller did as a graduate student in Robert Urey's chemistry lab at the University of Chicago in the 1950s, it has been notoriously easy to produce not just amino acids but chains of amino acids as bits of primordial protein, simply by passing a spark (simulating lightning) through a closed apparatus containing methane, ammonia, hydrogen, and water, but no oxygen—conditions such as would have existed on Earth four billion years ago, before the presence of oxygen in appreciable quantities in Earth's atmosphere. Protein-like molecules form rather readily, experiments show, and some theorists think that something like reverse transcriptase may have created the first nucleic acids from the structure of spontaneously formed proteins in this, the most elemental form of the dance between economics and reproduction (replication) in the systems we call life.

So far, it looks as if both economics and reproduction, intertwined, yet asymmetrically related to each other as they are, are both essential ingredients of life. In the short term—meaning the moment-by-moment survival of each and every organism—economics is way more important; in the long term, without reproduction, life would have ceased billions of years ago. And no "purpose" to life is manifest in this description: organisms are just there, processing energy and occasionally making babies.

## ENTER COMPETITION AND NATURAL SELECTION

So far, so good. It is probably safe to say that no biologist would have much of a problem with the description of economics and reproduction just outlined. But the argument ratchets up a notch or two when competition—for economic resources, for mates—enters the picture.

Darwin opened the first chapter of his *On the Origin of Species* (1859) with a discussion of pigeon breeding. People clearly had been modifying the features of animals and plants for a long while—increasing milk yields in cattle, producing different "breeds" of dogs and cats, and creating a spectrum of color patterns in pigeons. Darwin studied the process and learned that breeders, who see variation in the features they want to enhance, allow only those few offspring who show the best development of that feature—that color, that coat length, that milk yield—to mate with others similarly endowed.

Darwin knew, too, that animals and plants in the wild also show variation. He saw variation within local populations, and variation over the geographic distributions of species as well. And he wondered what, in the natural world, might be emulating the artificial selection wrought by breeders.

Darwin had read Thomas Malthus's *Essay on Population*[3] (1798) and absorbed the message that populations are limited by economic resources. Malthus was concerned with human populations, but Darwin saw that the principle must hold true for all natural populations as well.

All the ingredients were in place for Darwin's great insight: natural selection. Given that more organisms are born each generation than can possibly survive and reproduce (assuming, as is reasonable, that the population is already at or near its resource-limited size, and given that sexually reproducing organisms in the wild typically produce more than one offspring apiece), Darwin reasoned that *those best at prevailing in the economic game of life would on the average be more likely to survive to reproductive age—and actually to reproduce.* And because organisms tend to resemble their parents (for reasons Darwin knew not—the modern science of genetics was still decades away from its own birth), those organisms that succeeded best in the economic game of life would tend to pass on the traits that conferred that success in the first place. If, however, the conditions of life (as Darwin put it) changed over time, perhaps other variants would then be at an advantage, and the frequencies and even the natures of the traits would be "naturally selected."

Knowing how important his concept was, Darwin summarized the essence of natural selection in a foreword to the *Origin*, before he even embarked on his evidence and argument deriving and supporting the idea. He wrote,

> As many more individuals of each species are born than can possibly survive; and as, consequently, there is a frequently recurring struggle for existence, it follows that any being, if it vary however slightly in any manner profitable to itself, under the complex and sometimes varying conditions of life, will have a better chance of surviving, and thus be *naturally selected*. From the strong principle of inheritance, any selected variety

will tend to propagate its new and modified form. (*Origin of Species*, 1859, p. 5)

Darwin plainly means that how well an organism plays in the economic game of life will determine its survival—and thus also how well it will fare in the reproductive arena. Competition for food and other aspects of economic survival have, as a side effect, implications for reproductive success. Pure and simple.

Darwin himself, though, detected an exception to his rule that relative economic success spawns relative reproductive success. Organisms compete for food and other resources, to be sure, but they also can be found competing for mates. Some organisms are simply better at finding mates and at mating successfully—without any regard to how more efficient they are at making a living or otherwise surviving in a harsh world of finite exploitable resources. For this additional arena of competition leading to differential reproductive success, Darwin coined the term "sexual selection," publishing the idea in his 1871 volume *The Descent of Man*; to Darwin, sexual selection is "the advantage which certain individuals have over other individuals of the same sex and species, in exclusive relation to reproduction" (p. 256).

So there were, by 1871, two different sources of bias in reproductive success accepted in evolutionary biology: *natural selection*, the bias that relative economic success has on reproduction, and *sexual selection*, the bias on reproductive success stemming from purely reproductive features and behaviors. But the notion of sexual selection has, until relatively recently, languished far behind the idea of natural selection in the minds of most evolutionary biologists. That is because, in a sense, evolutionary biologists have not wanted to see the importance of the distinction; indeed, if (as selfish-gene thinking would have it) *all* competition, whether for mates or for resources, is *really* just competition for reproductive success, it makes little difference whether the competition is directly for mates or for food. To a modern ultra-Darwinian, it all boils down to the same thing.

I have already suggested that discoveries in molecular biology set off a chain of reactions among evolutionary geneticists that led to the notion of the selfish gene. One of these reactions was a desire to see evolution through natural selection as an active, creative process rather than as the simple, passive recording of what worked better than what, envisioned originally by Darwin, Wallace, and succeeding early evolutionary biologists. But there were prior movements in this direction—most important among them the evolution of the very term "fitness" in evolutionary thinking.

"Fitness" in common parlance has always meant "robust health," especially in the sense of "being in shape." Though common parlance still sees natural selection as "survival of the fittest," few know that it was the early sociologist Herbert Spencer, rather than his contemporary Charles Darwin, who coined the catchy phrase (and it was actually the poet Tennyson who anticipated the Darwinian vision as "Nature red in tooth and claw"). Creationists still occasionally tilt at the "survival of the fittest" windmill, claiming it is a self-defeating tautology (Who are the fit? Those that survive! Who survives? Those that are fit!). But Darwin was merely saying that, given variation in populations, and given a world where more offspring are born than can possibly survive and reproduce, those that are best at making a living ("fit"—meaning "hale and hearty") will, on the average, be the ones that will also reproduce. In other words, natural selection is more than the "survival" of the fittest; it is also the reproductive success of those that survive. But reproduction is a side effect of fitness in this original sense: the fittest do better in economic life and, as a consequence, tend to make relatively more babies.

Ever since the 1930s, however, in a stunning reversal of its original meaning, "fitness" has come to mean the probability that an organism will reproduce successfully—relative to the other members of its population. Here "fitness" has become a virtual synonym of "reproductive success," and the "hale and hearty" part of the definition is tacitly assumed. Sure, any ultra-Darwinian will say, that probability may be dependent upon how well the organism does in the competitive game of its economic life; for example, if we are talking about adaptations for predation, such as sharp teeth and keen eyesight in foxes, variation in hunting success in a fox population will be the determining factor underlying that spectrum of reproductive success.

But that old, clear-cut Darwinian distinction between economic and reproductive behavior, with natural selection being the effect of relative economic success on reproductive success, is muddied with the modern notion of fitness.[4] It is very much as if two clearly separate concepts have been jammed together— success in the act of living in the seething world of matter and energy flow, and success in making babies. And forcing together these two themes in the revised definition of "fitness" has contributed mightily to that desired effect of making the evolutionary process look and feel more like a bona fide scientific process, like the formation of a water molecule when two hydrogen atoms combine with an oxygen atom. Selfish-gene imagery (for that is all it is) is the extreme end product of this confusion of competition, economics, and reproduction in the biological world—at the huge cost to conceptual clarity.

## GENETIC IMPERATIVES:
## FOR AND AGAINST THE SELFISH GENE

But it must be more than mere computational convenience, or the sort of "physics envy" I suggested beset evolutionary geneticists once molecular biology came of age and started eating the rest of biology's lunch, that makes even the most level-headed evolutionary geneticist agree that hens are an egg's way of making another egg—or that the instructions for building a (biological) system are indeed more important than the system itself.

True, there is an innate "drive" to reproduce. It is also true that some organisms, like mayflies and salmon, reproduce only at the end of their lives, perhaps giving the impression that their entire lives were lived to get them to that final, crowning achievement of trying to leave offspring. But for any one kind of sexually reproducing organism that leaves reproduction to the end of the life cycle, there are hundreds that get reproduction out of the way early or, like us, take care of it in our younger, generally most robust adult years. But the mere fact of a "drive" to reproduce does not automatically suggest that reproduction is somehow more fundamental, more important than the drive to eat and simply stay alive.

Another factor at work here—leading to selfish-gene thinking and to the sense that genes are far more important than the organisms that carry and spread them—is the (potential) immortality of the genes, in sharp contrast to the mortality of all organisms.[5] Genes as physical entities, of course, are as mortal as any other part of an organism: *it is, rather, the information they contain that is potentially immortal.* And that there is a fundamental sameness to the DNA and RNA permeating all forms of still-existing life is ample testimony to the staying power of genetic information: some of it has already lasted over 3.5 billion years.

This staying power of genes, and the ephemeral life of organisms as mere vehicles for carrying and transmitting these "immortal coils," directly underlies much of selfish-gene thinking. For example, Nobel laureate Francis Crick's book *Life Itself* (1981) contains an arresting passage proposing an extraterrestrial origin for life on Earth. Crick argues that the origin of life is an inherently improbable biochemical event, thus unlikely to have happened on Earth. Without acknowledging that he simply sets the biochemical problem back to another time and place, Crick speculates that an advanced civilization somewhere out in the cosmos, facing doom and unable to save itself—its own genes—elects instead to save its bacteria and *their* genes, which crash-landed here on Earth between 3.5 and 4 billion years ago.

Although Crick's scenario at first glance seems the ultimate *selfless* genetic act (i.e., the simple desire to save life in whatever form, though it not be one's own), it is really the most extreme scenario of selfish-gene thinking I've ever encountered: we humans are descended from those primordial bacteria, and we indeed share genes with them. "Kin selection" and "reciprocal altruism" are notions applied by sociobiologists to explain the cooperation manifest in beehives, colonies, and social aggregations of vertebrate animals. But why stop there? We share almost 99 percent of our genes with chimpanzees, so one might extend the argument (or pare down Crick's fancies to a more local level) and predict that we would be more interested in the welfare of chimps and other great apes than of, say, lions, rats, fish, daisies, or mushrooms. And though people tend to relate more with other vertebrates than with fungi, and while many people are drawn to chimps because they remind them so much of themselves, truth be known that chimps hover on the verge of extinction through human agency as much as or more than most of the rest of the world's ten million other species.

## Is Life about Evolving, or Is Life about Living?

Evolutionary biologists—and I am one of them—take the evolution of life, and especially the processes that brought about 3.5 billion years of evolutionary history, very seriously. We tend to obsess about the details and argue endlessly over the nuances. Passion is a good thing, and it is very human to be deeply involved with what, after all, has proven to be one of the great ideas in Western intellectual history.

But sometimes this passion can be taken too far. Early in the twentieth century, anthropologists divided themselves into warring factions, including the "evolutionists" and the "structuralists." A structuralist felt that you could walk into a village, describe its contents and customs, figure out how everything worked, and thereby have understood the entire cultural system. An evolutionist, in contrast, said you could not understand a society unless you understood where it came from, what it grew out of—how it evolved.

It always seemed pretty obvious to me that it would be essential to have both a detailed account of the structure and function of a system *and* an understanding of its history. In fact, I still don't see how you can realistically have the one without the other; it might be possible to have a static description of what is there now, but how can you understand how something evolved if you don't have a detailed description of it, how it works and what its predecessors looked like?

Something very like this dilemma besets modern evolutionary biology. The notion of the selfish gene is an attempt to provide a functional description of biological dynamics, but it is a description of genetics, not of the structure, contents, and dynamic processes of the economic world—the world of ecosystems, where matter and energy flow between the inanimate and the living worlds, and between organisms of different species within the animate world. To my mind, selfish-gene evolutionary biology tries to explain the existence and evolution of a system—life—without providing an accurate description and analysis of that system.

Because an adaptation cannot be modified without differential reproductive success, ultra-Darwinian evolutionary biologists see reproductive success as the "final cause" of evolution; it is but a small step farther to see economic behavior as being really all about reproductive success—and even a smaller step to interpret life as fundamentally about the spread of genes, and to conclude that, ultimately, it is the genes themselves that are the driving force: genes control everything in life, and competition between genes drives evolution.

But this is a distorted description of life, and only a partially accurate description of how evolution works. Evolutionary biologists look at living systems and assume that the dynamics they observe—structure, functions, and behaviors—are there because they have evolved and surely will evolve some more. And in this, of course, they are right. They then conclude that the dynamics they observe at the moment are all *about* evolution. But they are not about evolution: they are about living life in the moment—about obtaining and processing energy, about avoiding being eaten and resisting disease, and, sometimes, about making babies.

Life is not about evolution. Life is about economics and reproduction—about the flow of matter and energy within and between organisms, and the information used by each generation for the production of new, replacement organisms. It is the interplay between these two components that, over generations, produces the patterns of stability and change in the information that we call evolution. Darwin was right in the first instance: environments change, new opportunities arise, new "varieties" appear (ultimately through mutation, we now know, so the traits of organisms are bound to change). There is no active "evolutionary process" per se. Evolution is simply a fallout—a record of what worked better than what in a finite and changeable world, a record that may change from generation to generation (though, as we'll soon see, species often remain adaptively very stable for millions of years). There is a deep distinction between the structure and the function of a system, and the origins and the further modifications of that system. In biological evolution, the two are related through a complex feedback loop: structural innovations arise from the genes, and how well they function has a side effect on reproductive success, altering genetic frequencies. The functions are not all about reproduction, still less about the further evolution of the system. The system—the organism, the ecosystem—is just there.

And though, in local populations, there can be competition expressly over reproductive success (for mates; or among sperms—i.e., true sexual selection), there is no cosmic need for genes to shoulder one another aside. Indeed, from this point of view, it is hard to see the "benefit" individual organisms supposedly derive from seeing relatively more copies of their genes left to the next generation. It is even harder to see (and a bit ludicrous to believe) how bits of information encoded as genes on chromosomes can in any meaningful sense "care" about leaving relatively more copies of themselves to the next generation.[6] The selfish gene is a metaphor predicated on the assumption that the purpose of life is to reproduce. *Au contraire:* if life can be said to have a purpose at all, it is simply to live.

The image of the selfish gene does no particular violence to the analysis of small-scale, localized evolutionary dynamics within populations, especially issues of sexual selection (from peacock plumes down to the competition among pollen grains or sperm cells to achieve fertilization). It causes no damage, either, in the analysis of the structure, function, and evolution of social systems—to the extent, in any case, that social systems are "reproductive cooperatives" (and they are often that, but also a good deal more than that, as we'll soon see).

But in obscuring the interactivity between the largely separate pursuits of an economic life and a reproductive life of most organisms, the concept of the selfish gene has been a setback for evolutionary biology. In the suggestion that the core of the evolutionary process is the "forcing" for stability and change coming from the fallout between genes to leave more copies of themselves to the next generation (or even the less extreme idea that competition for reproductive success is the ultimate cause of evolution), the importance of the environment in evolutionary history has been sharply undervalued (indeed, largely unexamined); the structure and functions of ecosystems, based in reality on the flow of matter and energy within and between components, are even now increasingly being seen, instead, as manifestations of sometimes cooperating, sometimes rival adaptations to spread genes. Worse, we have come to think that everything, including human behavior, is really about the spread of (our) genes.

So there we have it: we can either accept the proposition that the game of life reduces to the spread of genes, and that everything else organisms do is mere prelude or necessary positioning for those fateful reproductive episodes, or we can look at life, on a moment-by-moment basis, and see it as an interplay, a dance, between economics (matter–energy transfer) and information—a dance that depends equally on the two partners. These are indeed competing assumptions, perspectives on the fundamental nature of biological systems and what it is to be alive. There is no simple way to choose between the two—to show that one is false, the other true. Yet which way we jump on this chicken-and-egg issue determines to a large degree what we think species, ecosystems, and social systems are, and thus it ultimately informs our thinking about what manner of beast we humans are—and why we have sex. And why there is such a thing as sex in the first place.

## The Problem with Sex

Sex is a thorny issue for nearly everyone, and it is safe to assume that most humans have experienced sexual problems at one time or another. But evolutionary biologists, perhaps uniquely, have a problem with sex on a purely theoretical plane.[7]

The problem is this: sex is a horribly inefficient way of passing genes along. Every human child is a blend of 50 percent genes from the mother and 50 percent from the father. Asexual organisms, on the other hand, pass 100 percent of their genes along to each descendant. So the deeper question is not why *humans* have sex but why anyone—any species, from amoebae to elephants, mushrooms to redwood trees—has sex in the first place. If life and the evolutionary process really are all about spreading genes, why is there sex at all? Why didn't the world stay asexual? And why, given the origin of sex, didn't parthenogenesis (production of offspring from a single, maternal diploid parent) drive sexual reproduction off the game-of-life board?

Look at it another way: much of the rhetoric in evolutionary biology in the past fifty years was devoted to getting rid of the older, vague notion that evolution is "for the good of the species." Biologists like George Williams were particularly effective in arguing against such ideas, showing that evolution in the vast majority of cases must involve the good of the individual (as such, his argument was a direct forerunner of Dawkins's selfish-gene metaphor). Older explanations of the "good" of sex revolved around the potential for revitalizing each generation with new genetic combinations—good for organisms, and perhaps even for the greater good of the species, since it would provide the genetic variability for further evolution to take place. Williams convinced most evolutionary biologists that talk of the "good of the species," and of future evolution, doesn't hold up to close scrutiny: natural selection can "see" only what works best within each generation—and that means what works best for individual organisms.

Biologists have assumed that sex must provide some benefits to individual organisms—must be good for *something*—so that sex will be kept even at the "cost" of 50 percent lower efficiency in spreading genes. Unsurprisingly, a range of solutions to this apparent dilemma has been proposed. Some of the more promising involve the idea of repair: that errors (copying errors—"mutations") arising on a gene on one chromosome might be repaired by the "correct" version on the other chromosome, only possible when each gene occurs on both of a paired set of chromosomes—the "diploid" situation of sexually reproducing organisms. But it is striking that the main point of agreement on the issue in ultra-Darwinian circles over the past quarter century is that there *is* a problem at all—that sexual reproduction cuts in half the rate at which individuals can spread their genes, thus presenting a profound cost to them. The problem remains unsolved.

But consider this: there is a problem only if we assume that the struggle for existence is really all about the spread of an individual organism's genes. If, instead, there is no *competitive* urge to spread genes (whether arising from organism urges or more deeply from the individual genes themselves)—if there is no reason to suppose that the transmission of this bit of information rather than that bit is what really underlies the economic behavior of organisms as they search for food and otherwise live their lives, then sexual reproduction is no more costly to the individual than asexual reproduction.[8] Yes, the business of reproducing in some kinds of organisms can be overtly competitive: not, perhaps, among marine worms, sea urchins, and oysters, which simply send billions of eggs and sperm into the surrounding water to take their chances on fertilization, producing larvae, and having those larvae settle on the substrate, metamorphose, and grow up to be adult worms, sea urchins, and oysters. But certainly there is competition among, say, elks or peacocks—or even humans—for mates.

It turns out that sexual reproduction—for whatever direct benefits it may or may not have for individual organisms—creates communities of organisms that we call species. As we shall soon see, species are very stable entities, amazingly impervious to extinction. Species are much tougher to dislodge than clonal colonies of genetically identical organisms, and the staying power and genetic conservatism of species is the real reason why sexual reproduction is by far the method of choice in the macroscopic world of multicellular fungi, plants, and animals.

So people have sex in part because it is their evolutionary heritage. But there are many more consequences to this division of labor between economics and reproduction—or the realms of matter–energy transfer and the genetic information, and the circular interactions between them—that shed still more light on why people have sex.

# THREE

# The Natural Economy

M ale braggarts to the contrary notwithstanding, no animals, and certainly no humans, spend most of their time having sex. Even if the late basketball star Wilt Chamberlain really did have sex (as he claimed) with ten thousand women in his (very busy!) lifetime, he nonetheless spent most of his time every day in the far more mundane pursuits of material existence: eating, sleeping, going from place to place, and, of course, playing basketball, which he did at an awesome level. Even counting all the time spent in courtship behavior, most of an organism's life is focused on simple survival—that list of essential activities we've seen before. Even if we take the gene's eye view and claim that life really boils down to spreading our genes, we must concede that it takes energy to have sex.

Then, too, with the exception of only a few of our closer relatives (such as pygmy chimps, or bonobos), most animals have sex only at certain times of the year; most mammals (like elks and elephants), usually only during a single mating season. For the rest of the time, animals are living their lives, seeking energy and nutrient sources, consuming them, warding off predators and disease, and otherwise coping with life's endless challenges. One's economic life predominates, and this applies to humans as well as to all other forms of life on the planet.

Take a walk in the woods in summertime, and you'll see the economic side of life everywhere. Things are deceptively quiet. At first it's tough to pick up any action at all. Yet there's a palpable sense that something is going on. But then you remember that, in the presence of light, plants photosynthesize. In a biochemical trick invented over three billion years ago, plants use chlorophyll to catalyze the synthesis of sugar from the simple and ubiquitous ingredients of carbon dioxide and water. Just standing there, making no sounds save the breezes soughing through their branches, plants are hard at work soaking up carbon dioxide, sucking water in through their roots, and silently making sugar. This most elemental of all economic processes lies at the very heart of all life on Earth: no photosynthesis, no growth of plants; no plants, no animal life; no us.

But then you hear a rustling in the leaves; it's a towhee, a little black, white, and russet bird kicking up the dead leaves in the undergrowth, looking for worms and grubs, which it speedily gobbles up. The grubs and worms were eating dead vegetation left over from last season, while their brethren caterpillars were busily munching the green leaves overhead. You probably won't see it, but that towhee is destined to be taken by a sharp-shinned hawk, a small yet lightning-quick predator that lives on small birds and mammals.

Energy seethes through every woodlot, every field, every pond. Every creature needs energy to live, and most animals spend most of their waking hours looking for food, either dining directly on plants or devouring other animals. Eating a leaf or another animal's body transfers materials into one's own body—materials that the body digests, absorbs, circulates, metabolizes—crucial energy and material for building and maintaining one's own body. Nature may look serene, but the incessant flow of energy is everywhere.

Most of that flow is between distinctly different species, as when a squirrel eats an acorn fallen from an oak tree. Species are groups of organisms whose members share behavioral and anatomical features that allow them to exchange genes. For animals, of course, this means that members of a species see members of the opposite sex as potential sex partners. Every little bit of habitat almost always has more than a single member of any given species in it: you seldom see only one squirrel, one daylily, or one white pine. And, by and large, though it is easy enough to catch them having sex, you never see members of the same species exchanging matter and energy: cannibalism is not unheard of, but is comparatively rare in your backyard. Things that eat other things tend, as the overwhelming rule, to eat members of *other* species.

## LIFE IN THE LOCAL ECOSYSTEM

But if species are mating clubs, what does this have to do with the economic side of life, the business of making a living? It turns out that birds of a feather flock together not just during the mating season. Members of the same species living in the same patch of woods have basically the same dietary requirements; they need similar amounts of water and are most comfortable in the same basic temperatures, day and night and through the seasons. In brief, they share the same basic economic needs, which is why the earliest biologists defined species simply as groups of similar animals or plants. We now know these organisms are so similar because they share a gene pool, one that is constantly being mixed and matched—shaken up, really—by the annual mating ritual. And that is why biologists have for half a century now seen species as reproductive communities, groups where sex and baby making go on among its individual members, but *not* (at least as the overwhelming rule) between members of different reproductive groups. Gray squirrels mate with gray squirrels, chipmunks with chipmunks—and human beings with other human beings (the exceptions of bestiality to the contrary notwithstanding—and in any case never leading to the production of offspring). Taking Weismann up a step, we note that some behaviors and anatomies evolved through natural selection (adaptations) are used for making a living and that others are used for sex. Species are defined by the shared behaviors and anatomical features of organisms for having sex, and not by their shared adaptations for making a living.

But, on the local level, members of the same species share a slew of economic needs. If energy isn't flowing between them, they are nonetheless constantly interacting. Sometimes they are competing for resources, since there is only so much energy available in a local system that any particular group is able to extract—meaning that, if you are a bobcat, and your normal prey is mice and rabbits, you will find only so many mice and rabbits living in the neighborhood. It's mostly limited energy supplies that determine how many of a given species can occupy a single place—the limiting factor that both Darwin and Wallace found critical to the very idea of natural selection. Darwin saw a world teeming with competition among members of each species, all of them vying for the limited energy resources available to them. And if it is far easier to catch male elks competing for mates than to measure how they compete for their forage, the world of elk food is nonetheless finite (as the need to feed elks in the wintertime around Jackson Hole, Wyoming, makes all too clear). This is Darwin's struggle for existence, one in which only the best adapted, strongest, and healthiest ("fittest") would tend to survive. As a side effect, the fit would be more successful in the game of mating and baby making.

But out-and-out competition is only part of the interactive story. I was once on safari in Botswana's Okavango swamp when a horrendous din broke out just a few yards off the road where we'd stopped to study some wild mistletoe. Our guide jumped behind the wheel and plunged us into the thorn scrub. There we saw the last moments of a short and violent confrontation between three wild dogs and five spotted hyenas. The dogs had just killed an impala, and now the hyenas were ganging up on the dogs and wrenching the last pieces of the impala carcass away from them. Had the dogs kept the carcass, they would have shared it according to the rules governing their pack. The hyenas, in contrast, started hooting and howling their maniacal cackles as they made off with a haunch here and a leg there, all vestiges of cooperation instantly dropped as they competed fiercely with one another for parts of the already dismembered impala carcass. No sharing here among the hyenas! It was a startling episode, one that showed both the competitive and the cooperative sides of economic life in two distinct species of African carnivore acted out in less than two minutes.

This snarling mass of competing wild dog and hyena flesh, both groups united by total cooperation until the outcome triggered competitive behavior among the victorious hyenas, may be far more dramatic than anything you are likely to see in your nearest woodlot. But it is utterly typical of the interactions among the members of local groups and their relations with other groups. This sort of collective impact on the system is what ecologists usually mean by the term "niche." Local bands of hyenas hunt, scavenge, and steal food from their competitors. They eat a wide range of prey, fear only lions, snakes, buffalo, and elephants as adults (though their young are, of course, much more highly vulnerable to predators), and can stare down or beat up most of their direct competitors.

Spotted hyenas need to drink water, and they can tolerate only the fairly narrow range of daily and seasonal temperature fluctuations typical of the tropics. This combination of needs and impacts adds up to the spotted hyena niche in the savannas and woodlands of eastern and southern Africa. Living as they can on a wide variety of foodstuffs and in a range of habitats, from wetter woodlands to nearly barren dry grasslands, spotted hyenas occupy a fairly broad niche, despite their restriction to the warm climes of the African tropics.

So, too, do the impalas, a common species of African antelope. Although impalas cannot live without ready access to drinking water, they are otherwise able to live just about everywhere over the African continent, subsisting on an impressive range of leaves and grasses. Not so their closest relatives: species of wildebeest, hartebeest, and the like, most of which inhabit much smaller ranges, can tolerate only a much more limited variety of conditions, and can eat only a few particular species of plants (mostly grasses). These are the narrow-niched species, and it is an arresting fact that such species are far more prone to extinction than their more loosely adapted, broad-niched brethren species. For putting all your economic eggs in one basket, as ecological specialists basically do, is to risk not being able to swing with the changes in climate and altered food supply that inevitably come along. Hold this thought, because humans for the most part are also ecological generalists, though we have altered the rules by which we play the ecological game—a theme we'll revisit in part 2.

## ECONOMICS ON A GRANDER SCALE

So that's the real world, constantly in motion, energy coursing throughout the local system. But no system is an island—no field, no woodlot, no pond, no ocean. When a brown snake eagle takes a cobra in the grasslands of Botswana, when Pel's fishing owl or an African fish eagle grabs dinner from the Okavango River, or when a lion comes from the grassland savanna into the riverine forest to chase a bushbuck, all is energy leakage across the porous boundaries of local ecosystems. We can understand the flow of energy within an obviously bounded system like a pond, but these systems are in no way closed. Not only does energy come in the form of primordial sunlight (and bounce out again when not captured in photosynthesis or temporarily absorbed by rocks, soils, and water), but energy in the form of organisms constantly coming into and going out of the system ensures that all local ecosystems are connected with one another, from the mangrove swamps of the Okavango Delta to its river and flood plains, its riverine forests, its grasslands, its mopane woodlands. All are connected, maybe by something as striking as a herd of elephants walking through, munching grasses and palm leaves, tearing bark from tree trunks, and leaving huge piles of dung behind them. Or maybe by the much more subtle comings and goings of birds, whose own droppings also fertilize and also transmit microbes—viruses, bacteria, fungi—that are in many ways the energetics backbone of all ecological systems.

Thus eating, living an economic life, has huge consequences.

Success at it might make possible success at mating, to be sure, but it really means that all organisms, together with other members of their own species in the same neighborhood, are cogs in an energetics wheel that is the local ecosystem. And these local systems are all connected, by the constant flow of material and energy between adjacent systems. The entire globe is linked up this way, as local ecosystems link up to regional systems, and regional systems, like east African savannas and the (encroaching southward) Sahel/Sahara desert systems to the north, are also linked through the simple flow of energy across their boundaries. All of life, in the moment-by-moment energetics sense, is hooked up into one grand, enormous, ultimately global system.

This grand global network of life is the biosphere. It is intimately connected with the atmosphere, hydrosphere (the waters of the planet), and lithosphere (rocks and soils)—all of which systems are also interacting ceaselessly. Climate, for example, is deeply connected to the circulation patterns of the oceans— the various warm and cold currents ultimately driven by solar energy and the topography of the lithosphere—the disposition of the oceanic and continental plates.

James Lovelock has called this intricate global system "Gaia," borrowing the ancient Greek word for "earth." Some Gaian enthusiasts have gone too far, by suggesting that Earth is really like a gigantic "superorganism"—with the lithosphere, for example, serving as the skeleton for the thin veneer of life processing energy interconnected over virtually the entire surface of the globe. As the biologist Lynn Margulis points out, though, no true organism eats its own effluvia. The endless global cycle of life, death, decay (largely through the action of bacteria and fungi), and renewal is very different from the births and deaths of individual organisms. Yet the essential truth remains: there exists a prodigious, worldwide flow of energy hooking up the physical realm with all of life, no matter what we might choose to call it.[1]

Ecologists are the ones who study these systems. The search for connections between ecology and evolution over the years has not been smooth. Intuitively, everyone feels the two fields must be connected, but what those connections might look like has remained elusive. Unsurprisingly, some gene-centered evolutionary biologists think that the structure and function of ecosystems will come to be understood in terms of differential success that organisms within populations within an ecosystem will have in spreading their genes.[2] The search for connections ends up with evolutionary biology's eating ecology's lunch.

Ecosystems, of whatever size, are composed of groups of interacting organisms; matter–energy transfer is rampant and constant. The organization—structure and function—of an ecosystem, complex and difficult to analyze as even the simplest of them may be, is all about energy transfer. Sure, the adaptations used by each and every different type of organism in the system is the product of a long evolutionary process. And those adaptations are also bound to be modified over time as the physical environment and other biological components of the system change. But trying to describe the dynamics that define and cohere an ecosystem in terms of differential genetic success is like trying to understand the chemical bonding of oxygen and hydrogen atoms to form a molecule of water in terms of subatomic particles. Worse, actually, because, at least in principle, subatomic particle behavior gives rise to the behavior of atoms. But genes, important as they are, are not involved in moment-by-moment energetics—of an organism's body, or between one organism and another. Explaining ecosystems in terms of active competition for reproductive success is like trying to use the rules for grammatical structure to describe a baseball game.

Reducing one field—ecology—into the terms of the other when the very nature of the processes within each are so fundamentally different is a fool's errand.

Yet connections there are between the realm of matter–energy transfer—the economics of living systems—and the storage and utilization of the information pertaining to the construction of individual organisms, that is, genes. Natural selection as originally conceived offers a prime example: a circular feedback system, where genes make organisms, organisms live their lives, and, as a side effect, how well they fare in the economic sector biases the composition of the gene pool used to make the next set of offspring, the new players in the game of life. The connections are always through the organisms, which are simultaneously players in the economic game of life and (usually) in the business of making babies. How this interaction between economics and having sex plays out in the evolutionary history of life is dramatic—and the subject of chapter 5.

So the mere action of staying alive—of processing energy and nutrients to develop, grow, and simply to live—automatically creates a magnificent economic (or ecological) hierarchy: every organism needs energy, and so finds itself interacting locally with other members of its own species; these local groups (often called avatars) in turn play specific roles in ("have niches"), and thus are parts of, local ecosystems; local ecosystems, in turn, are linked up with neighboring systems as energy flows across the boundaries; and these regional systems are all hooked up, ultimately, to the biosphere. I have shown this ecological hierarchy in diagrammatic fashion in figure 1.

## Evolutionary Hierarchy

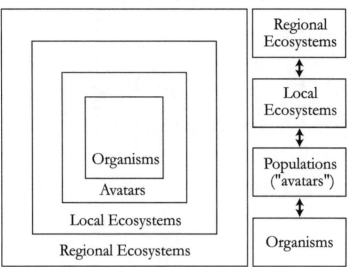

FIGURE 1. Two ways of looking at the ecological hierarchy. *Left:* Regional ecosystems contain local ecosystems, which are made up of populations (avatars), themselves composed of individual organisms of a single species. This arrangement emphasizes the inclusive, "nested sets" aspect of the ecological hierarchy. *Right:* The ecological hierarchy seen dynamically, with upward and downward economic interactions (matter–energy transfer) between levels symbolized by dark arrows. (All illustrations by Steve Thurston.)

# FOUR

# The Consequences of Baby Making

L ions don't eat while mating, one big reason why their
tempers are so short as they snarl their way through a
prodigious number of couplings in the two or three
days male and female stay together. They rely on their storage
of energy, their fat reserves, to keep them going and to power
the frenetic humping that tourists on safari are so fond of ogling
and photographing.

Mating in the African bush, in the crannies of coral reefs, or
on the ice floes of the Arctic truly is all about making babies.
Social structure might determine who gets to copulate (meaning, for the most part, which males, as in the dominance hierarchies of elephant seals, where only a few enormous bulls in their
physical, economic prime, get to service whole herds of cows).
It doesn't work the other way around. Having sex does not
determine social structure, at least not among lions, moray eels,

or polar bears. Sex is not about economics with all these disparate species; it is about making babies, pure and simple.

Throw some yeast into a cup of water with some sugar, and watch what happens. The surface becomes milky with small bubbles, the carbon dioxide excreted by these single-celled fungi as they metabolize the sugar. Yeast cells mobilize instantly to take that sugar in and use it for their cellular needs; but yeast cells do not so much grow as divide, and divide and divide. In other words, probably the most obvious thing that yeast cells do when faced with the cushy environment provided by water and sugar is *reproduce*.[1] Yet, though the distinction between economic and reproductive sides of life is not so readily apparent in a yeast colony as it might be in a maple tree (where flowers are the locus for pollination and seed production and the rest of the tree is devoted to nonreproductive functions), the distinction is nonetheless there. It is only those cells that can successfully take up and metabolize the sugar molecules that can divide. And those that do it best divide more quickly and leave more descendant yeast cells behind: selection once again in action.

## What Is a Species?

Animals face certain practical necessities that must be met if they are going to reproduce successfully. First off, they have to find a mate, and, as most humans will attest, this is by no means a cut-and-dried matter. How do they locate a mate, and how do they tell what kind of animal is even an appropriate potential mate? The answer varies, of course, with the nature of the beast. Lions recognize one another by scent and visual cues, and males and females know one another by particular subsets of those scents and visual cues. But how do corals, rooted as they are to the substrate, manage to find mates in the sexual phase of their reproductive lives? Corals can also reproduce simply by asexual budding, where a new polyp literally "buds off" from a parental polyp, the main way corals and many other marine backbone-less ("invertebrate") animals form their colonies.[2] By the same token, how does a maple tree "find" prospective mates? How does a female eastern bluebird manage to find and choose a male eastern bluebird to make babies with, and not be tempted to mate with a house wren or an eastern phoebe?

Every species has a "mate recognition system."[3] It might be chemical: corals, sea urchins, and many other marine creatures simply shed their eggs and sperm into the oceanic waters that bathe them; chance encounters of eggs and sperm of many different species lead nowhere because the eggs and sperm chemically recognize only their counterparts from the same species. A marine annelid worm's sperm cannot fertilize a sea urchin egg, or the egg of another species of worm, for that matter. Even in bizarre cases where external cues break down and individuals of remotely related species of mammals may couple (such as the storied "animal acts" in the colorful history of bordellos— women and police dogs, for example), conception is impossible because of the incompatibility not only of the genetic makeup of egg and sperm but also of the chemical incongruities that prevent the sperm and egg from "recognizing" one another.

The obvious competition for mates among rutting elks and strutting peacocks is one of the main reasons why biologists are drawn to the notion that life really boils down, in the end, to reproducing. Darwin's "struggle for existence," which for him meant primarily competition for energy resources, is difficult to observe in the wild. It is much easier to point to mammals and birds that openly compete for mates. But what, then, of the crapshoot of wind-blown pollination and the profligate (and often synchronic) shedding of gametes that the bulk of marine creatures use? It's very hard to think of thousands of oysters shedding their eggs and sperm as indulging in competition for reproductive success in any sort of realistic manner. The counterclaim might be that the healthiest, most robust oysters in fact shed more gametes than their less robust neighbors, but that is untested speculation that need not, in any case, be true. Even barnacles, which are permanently rooted crustaceans and which indulge in a more direct form of fertilization, are still locked into a game of pure chance: the males of some barnacle species are equipped with enormous penises that, compared with overall body size, would be the envy of *any* guy—yet even these prodigious organs are finite in length, and the barnacles can inseminate only the neighbors within their reach. And who is to say how "fit" these neighbors are—as barnacle larvae settle out at random from the seawater above, and metamorphose and survive to grow into adults only if they happen to land on suitable substrate on which they can attach and if there are sufficient nutrients to be strained by the gills from the surrounding seawater? Barnacles have no control over who their barnacle neighbors are.

Obsession with the image of the selfish gene has led to numerous reports of competitive races between rival sperms to get to a ripe egg—or of pollen (the plant equivalent of sperm) to reach the ovule. Though some of these studies may indeed hold up, a more recent report has shown that there is a sort of cooperative "buddy system" among human sperm cells as they make the difficult, dangerous journey (through an oddly chemically hostile environment) to make it through a woman's cervix and there to seek out and successfully fertilize the (as a rule) single ripe egg that is present for a few days each month in reproductively active females. Generally, once a sperm cell penetrates the outer wall of the egg, there is a chemical change in the egg's outer wall that prevents more sperms (now the "losers," if the imagery remains in the competitive mode) from entering the egg. All the other sperm cells are destined to die and be flushed away. But how interesting this notion that sperms can actually band together in the hope that one of them will win the lottery![4] On a higher level, whales will sometimes lend the weight of their bodies to help copulating couples bring their mating to fruition.

So, overt, no-holds-barred competition in the mating arena is, in the last analysis, relatively rare. In contrast, mate recognition systems are ubiquitous. Recognition systems may be as subtle as the chemical compatibilities between egg and sperm, or as robust as the song of male birds that is unique to each species. Male songbirds frequently also have a distinct color pattern often lost after the mating season. Gorgeous male indigo buntings, for example, take on more of the look of the drab brown females in the fall. When mating is no longer the order of the day, priorities shift back to avoiding being eaten, and drab brown works a lot better as camouflage than the iridescent blue that males wear to attract a female during the mating season. Often, males show up first on the breeding territories, stake out a territory (usually constantly defended against intruding males), and start singing and displaying their plumage to attract a mate. Female birds know the song from infancy (their fathers sang it) and recognize the plumage.[5] Though heartiness of song and finery in plumage are generally assumed to be markers of the genetic health of the prospective father, females are known to take a good look at the territory the male has staked out. From an economic point of view, parents not only have to eat but also to feed their brood until the kids can fend for themselves.

That's what species *are*: breeding collectives. Species consist of all the organisms that share a mate recognition system.

Change that system, so that organisms no longer recognize one another, and you have two species where once you had only one. "Speciation" minimally entails merely a change in the mating systems of groups of organisms, which can otherwise be virtually identical. North American damselfly ("darning needle") species, for example, often differ only in the configuration of the male and female copulating parts and are otherwise anatomically and ecologically indistinguishable.

On the other hand, as Darwin pointed out long ago, most of the species represented in your backyard are readily distinguishable from one another. Each usually has relatives living elsewhere that look much more nearly alike. Take chickadees: present in most backyards throughout the Northern Hemisphere, chickadees form a complex of related species that are often hard to tell apart. Here in the northeastern United States (where I can hear the black-capped chickadee, *Poecile atricapilla*, calling as I write), the closest local relative is the equally vocal, but very different-looking, tufted titmouse *Baeolophus bicolor*. But somewhere in southern New Jersey, along the Mason-Dixon line (i.e., the southern border of Pennsylvania projected eastward across New Jersey to the Atlantic Ocean), the calls of the chickadees seem somehow speeded up, and the birds seem a little bit smaller. That's the Carolina chickadee, *Poecile carolinensis*, a species so similar to the northern black-capped that even experienced birders can be fooled. On the other hand, the chickadees on the higher peaks in the Adirondacks have brown caps and more russety sides, so they (the "boreal chickadee," *Poecile hudsonica*) are easier to tell apart from the black-capped species.

Black-capped and Carolina chickadees, like the indigo and lazuli buntings already encountered, learn their songs from older males—and, once again, sometimes the wrong song is learned, leading to hybridization along the line where the two species meet.[6] But, on the whole, a sharp more or less east–west line divides the two species—the black-capped always to the north, the Carolina to the south. No matter how similar in appearance, song, and even apparent ecology, the Carolina chickadee is what you find in Virginia, while the black-capped, and only the black-capped, chickadee is what you get in Michigan. And that means that, despite the sporadic hybridization where the two species meet, very little, if any, of the Carolina chickadee genome—to the extent that it differs from the black-capped chickadee genome—is spreading into the black-capped species, and vice versa.

These two closely related chickadee species tell us a lot about what species are and how they come to be. Geography is the key to understanding how populations of sexually reproducing animals and plants can diverge, forming two separate species from a single ancestral species. When species are spread out over broad areas, as they often are, far-flung populations are not in direct contact with one another. Chickadees tend to live in the same area year round, though falling back from extreme cold or snowy conditions a bit perhaps. And while adjacent populations (or "demes"—see below) are often in contact, so that potentially, in network-like fashion, there is reproductive continuity between all parts of a species throughout its range, genetic divergence is nonetheless nearly inevitable over large tracts of a species' distribution.

Humans provide a great example of a far-flung (global!) species that has visibly—and genetically—diverged over the past 100,000–200,000 years. But, just as the black-capped chickadees in Maine would almost surely "recognize" and successfully mate with those living east of the Rockies in Colorado, the human mate recognition system remains intact throughout the world. The age of exploration and conquest makes it abundantly clear that humans of all ethnicities enthusiastically hop into bed with one another, often making babies in the process. Though the entire subject of "race" is of course loaded, it is still obviously true that the differentiation of peoples genetically began in the same way that it did in all far-flung species: by the accumulation of differences (some through natural selection, such as the sickle-cell anemia gene—see chapter 7) or by the simple accumulation of mutations that did not occur in other populations and other random events in geographically disjunct populations.[7]

But *Homo sapiens*, like the black-capped chickadee, remains a single species—no matter how much diversity has sprung up within the ranks. No matter how widespread, no matter how diversified ecologically or even reproductively, a species becomes, there is no automatic threshold that sunders an ancestral species into two or more daughter species. Something more has to happen. It is becoming increasingly clear that "something more" usually involves environmental events—events like prolonged cold snaps, or even impacts of comets and asteroids with Earth. Something, at any rate, that changes the distributions of species and isolates populations so thoroughly that these populations diverge to the point where interbreeding is no longer an option—should they survive and eventually make contact with populations of what used to be the same species.

What, for example, could explain the curious line of demarcation between black-capped chickadees and their southern Carolina chickadee cousins? Though I am no expert on bird speciation, that east–west line separating these two chickadee species looks to me suspiciously close to the line marking the southernmost extent of continental glaciers during the Ice Age. Ice receded the last time just over eighteen thousand years ago from its southernmost limit, and the vast oscillations in climate, habitat, and distributions of forests are the most likely factors that triggered the split between black-capped and Carolina chickadees. Similar as they are in song, plumage, and even general ecological behavior, these two species differ at the very least in their ability to survive long, bitterly cold winter nights.

## SPECIES: NATURE'S CONSERVATIVES

We'll get back to environmental triggers of speciation, and to evolutionary change generally, in the next chapter, when we begin to look at the interaction between the worlds of economics and sex. Here it is, in a sense, the opposite aspect of species that commands our attention: their incredible conservatism, their downright resistance to change, and their consummate ability to last for prodigious periods of time. Years ago, Steve Gould and I dubbed this resistance to change "stasis"—and claimed that it is a truly general feature, a mark of the vast majority of species that have ever graced the planet.[8]

Stasis is important, if only because it shows that evolutionary change—at least of the sort that shows up in bones, teeth, and shells of readily fossilized organisms, the sort that usually has to do with the economic adaptations of organisms—is hard won. Indeed, in a world where we now recognize rampant short-term fluctuations in the genetic composition and structure of populations, it is remarkable that most species, variable though they may be from place to place, remain virtually the same over hundreds of thousands, and often millions, of years.

This amazing stability of species—this stasis—is a direct by-product of sex, geography, and environment. Because species are broken up into semi-independent demes, each with its slightly different environment, its different mutational history, its different selection regimes—in short, its own separate mini-evolutionary trajectory—there is little chance that selection will drive the entire species in any one direction for any length of time. Paleontologists see their species moving back and forth, with some change in a particular direction for a while, most of the time only to be reversed later on.

Species arise from sex: sex requires (at least one) partner, chosen from a local population (deme) of potential partners. Species are simply summations of all the demes that share a specific mate recognition system. Species are stable because their demes have semiseparate evolutionary adventures, and so it is difficult (not impossible!) to accumulate species-wide evolutionary change over long periods of time. Species are stable (and also resistant to extinction, for reasons explored in the next chapter). The mere existence of sex thus leads to the formation of extremely stable entities—species. This is a major, probably *the* major, reason why sex predominates over asexual reproduction—and yet another reason to question the assumption that evolution should favor asexual systems that maximize transmission of an individual's genes.

## LINKAGES: SPECIES AND THE LINNAEAN HIERARCHY

Species are packages of genetic information. They don't actually *do* anything, at least not in an active dynamic sense comparable to the energy exchanges between populations that hold ecosystems together.[9] Entire species don't interact with other species, the way their local populations do inside ecosystems. Species are not parts of dynamic economic systems and do not exchange matter and energy with other species. They are, instead, packages of genetic information, parts of a hierarchy of genetic information with roots reaching back to the very inception of life over 3.5 billion years ago.

Reproduction keeps species going, and the production of babies is essential for the economic side of life to continue, keeping ecosystems functioning and repairing damaged ecosystems as well. Squirrels need to be born each year to maintain the squirrel population; no baby squirrels means less food for foxes and Cooper's hawks—and then *they* will make fewer babies. The reproductive side of life means that individuals (by hooking up with another of the opposite sex) can make more individuals ("babies"). Species do something similar, just up one notch. Species, as we've seen, make more species. Some of them, in a sense, even "have sex": some groups of plants (pine trees, rose and berry bushes) are particularly adept at hybridizing. Sometimes an entirely new species, whose individuals look different from the parental species and cannot breed with members of either parental species, evolves from the combination of the genetic materials from two separate species. Among animals, though, such events are rare.

So these reproductive communities, these species, make more of themselves ("speciation"). The lineages created by the evolutionary process, lineages that extend back into the mists of geologic time, coalescing as you look ever deeper until all merge at the single ancestral bacterial cell 3.5 billion years ago, are strings of ancestral-descendant clones (among the bacteria and some other microbes) and species—after sex comes to dominate the reproductive scene.

There is an obvious look to the biological world. Some species, such as apes and humans, look more like each other than like, say, frogs. And it is clear that humans, apes, and other primates are mammals, all species with hair, mammary glands, a placenta (save in marsupials and a few others), and three middle-ear bones. Frog species, on the other hand, show a sort of family resemblance, too, and look a lot like toads; frogs and toads share a form of reproductive life (tied to water) and other features that make them all "amphibians." Yet mammals and amphibians (along with birds, reptiles, and various fishes) all share a backbone; they are "vertebrates." This is the Linnaean hierarchy, where every species is united with other similar species into the same "genus." Genera are united into families, families into orders, orders into classes, and so on. Darwin told us why. Those species that resemble one another most closely are one another's nearest relations. They have a relatively recent common ancestor not so far back in geologic time. Less similar species are more remotely related, and so on back, not quite ad infinitum, but as far as 3.5 billion years, the time when the bac-terium ancestral to all of life was respiring and dividing.

The reproductive buck stops at species: all the higher level groups of the Linnaean hierarchy are collections of related species. A family—say, the hominid family (Hominidae), of which our species, *Homo sapiens*, is a part—does not give birth to other families. But that's a lot of reproducing, all the same. Genes are copied—sometimes by RNA to make proteins, and sometimes by a simple process of division to make more genes. Animals copulate, and their sex cells ("gametes") fuse to make babies; copulating creates pools of mating partners, and the largest collective is a species. And species make more species. That's the other side of life—"more-making"—a hierarchy that matches the ecological hierarchy, but comes from more-making, not the acquisition of energy and living the economic side of life. We could, with some justification, call this the "reproduc-tive" hierarchy, but "genetic" is a more general term, and per-haps "evolutionary" is best of all (figure 2).

So, the split between economics and reproduction, a

dichotomy that pervades all of life on Earth, means that absolutely every organism, past, present, and future, belongs in two separate systems. It is a part of a local economic system—a population (avatar), itself a part of a local ecosystem, then progressively of larger, regional ecosystems—and a local mating system—a population (deme), which is part of a species, with species being parts of larger-scale groupings of the Linnaean hierarchy.

Question: What connects these two different worlds of the economic and evolutionary hierarchies? How, in other words, do the worlds of economics and sex, so important to the lives of all organisms, intersect?

# Evolutionary Hierarchy

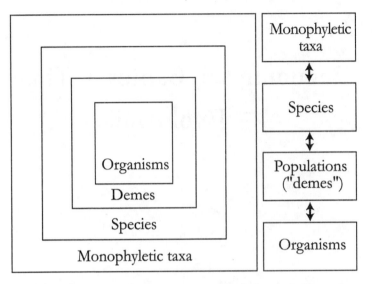

FIGURE 2. Two ways of looking at the evolutionary hierarchy. *Left:* Monophyletic taxa (such as phyla, classes, orders) are groups of related species, themselves made up of breeding populations (demes), composed of interbreeding organisms. This arrangement emphasizes the inclusive, "nested sets" aspect of the evolutionary hierarchy. *Right:* The evolutionary hierarchy seen dynamically, with more-making (reproduction, speciation) and extinction seen as elements of upward and downward interactions between levels.

# FIVE

# Economics + Babies + Time = Evolution

Natural selection, as we have seen, connects the economic side with the reproductive side of any organism's life. But there are other connections between the biological worlds of economics and reproduction, between the business of eating and the chance for having sex. If we set the two hierarchies side by side (figure 3), we see the connections immediately, within the bodies of individuals and within local populations.

A comparison of the two hierarchies in figure 3 really tells us how evolution works. Ultra-Darwinian evolutionary biologists like to think that genetic competition drives evolution, and creates ecosystems to boot. Everything in the twin hierarchies of life, they say, falls out of this constant urge for genes to outcompete one another to make it into the next generation. Sex rules the roost! In contrast, I say that ecosystems exist sim-

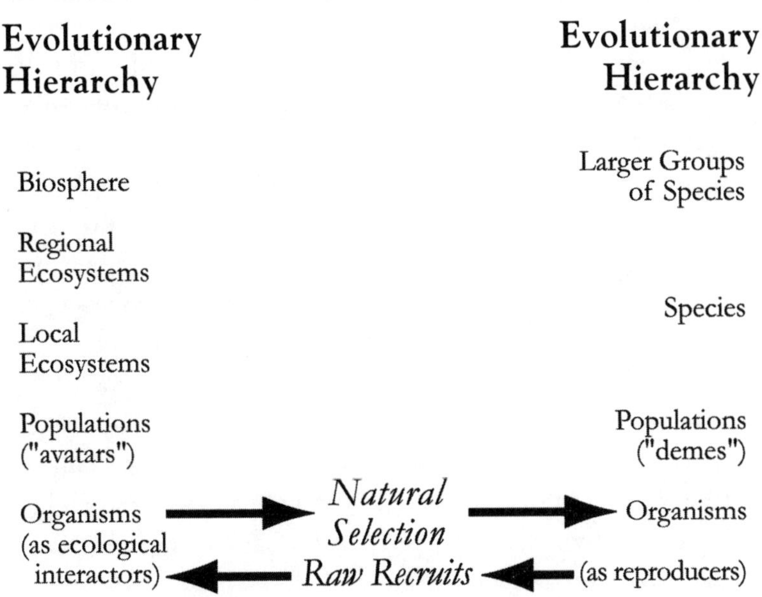

| Evolutionary Hierarchy | Evolutionary Hierarchy |
|---|---|
| Biosphere | Larger Groups of Species |
| Regional Ecosystems | |
| Local Ecosystems | Species |
| Populations ("avatars") | Populations ("demes") |

Organisms (as ecological interactors)  ⟶  *Natural Selection*  ⟶  Organisms

Organisms (as ecological interactors)  ⟵  *Raw Recruits*  ⟵  (as reproducers)

FIGURE 3. The ecological and evolutionary hierarchies seen side by side. Organisms are the only entities common to both hierarchical systems. Reproduction furnishes a constant supply of "raw recruits" to local ecosystems; how well organisms fare ecologically in turn has an effect on which organisms on the average will leave more offspring to the next succeeding generation—Darwin's original, and still the best, conceptualization of "natural selection."

ply because organisms need to eat, and the web of energy connecting all the populations of plants and animals, microbes and fungi, is what actually forms the ecosystem and keeps it running. The genetic side of the ledger is necessary only to keep a constant stream of new players in this economic game of life.

The simple, original evolutionary model saw species undergoing pretty much constant evolutionary change, as natural selection would continually be modifying the adaptations of organisms within all species to match inevitable and persistent environmental changes. Later evolutionary biologists (like Dobzhansky and Mayr) realized that speciation—the origin of new reproductively isolated communities from ancestral species—must play an important role in the evolutionary process. And when paleontologists began asserting the importance of stasis—the relative lack of discernible evolutionary change within vast stretches of most species' histories—it became clear that evolutionary history is no simple matter of natural selection tracking environmental change through time.

One fascinating thing about the fossil record (especially of the past half billion years, since complex forms of animal life have been living on the planet) is that a few simple patterns tend to recur over and over. Instead of thinking that each event—each speciation event or each mass extinction—is unique (which, in its details, is of course true), the fact that these kinds of events happen again and again in the history of life tells us much about how life actually evolves. These generic sorts of patterns—relatively few in number (I list the three I feel are most important here)—are like the fossilized results of experiments done in other sciences in the laboratory. True, we cannot set up carefully controlled conditions in advance of the "experiment." But when physicists studying the properties of subatomic particles chart the evanescent, yet revealing, trace of the passage of an elusive particle in the cloud chamber of a particle accelerator, they can deduce attributes such as charge and mass. And they can also decide whether their results confirm or overturn previously devised theoretical expectations.

We can do the same with patterns in biological nature: for example, patterns of geographic variation within and between species in the modern, living world eventually led to the elaboration of the notion of geographic ("allopatric") speciation. Patterns in the fossil record give us the added element of time in geography and have led to some exciting insights into the evolutionary process.

## THREE EVOLUTIONARY PATTERNS

Pattern no. 1 is stasis—that intransigent anatomical stability so characteristic of most species—as seen in Cambrian trilobites, Jurassic dinosaurs, Miocene horses, and even some Pliocene/Pleistocene hominids. All these sorts of animals of course evolved, but anatomical change from species to species tends to be concentrated in relatively brief spurts (compared with the longevity of the ancestral and descendant species). And that evolutionary change also seems to be associated with episodes of branching—splitting of lineages, speciation.

Speciation is pattern no. 2. Already encountered in chapter 4, speciation has been going on since the advent of sexual reproduction. Most evolutionary change seems to be associated with the fracturing of one extended reproductive group—a species— into two or more descendant species. Periodically, my Devonian trilobites, the dinosaurs of the Mesozoic, the horses of the Tertiary, and even the hominids of the Pliocene and Pleistocene would produce new species, many of them budding off from an ancestral species that persisted through time. More recently (and this is part of pattern no. 3), it has been found that ancestors often do *not* survive the rapid appearance of their descendants. For example, the Pleistocene hominid species *Australopithecus africanus*—an upright, bipedal hominid with a brain roughly 450 cc in size (comparable to that of a modern chimp)—lived from roughly 2.5 million years to 3 million years ago; it was replaced by not one but two hominid species and

may well have been ancestral to both. One probable descendant, the earliest discovered species of *Paranthropus*, had about
the same size brain—but a massive skull equipped with huge
grinding teeth, suitable for life as a nut-cracking, tuber-chewing
herbivore. The other lineage in all likelihood descended from
*A. africanus* was the earliest toolmaker—*Homo habilis*, equipped
with a far larger brain than its ancestor.

But it is pattern no. 3 that really shows us where the action
is and how the reproductive and ecological hierarchical systems
resonate with one another to produce the evolutionary history
of life. It turns out that the give-and-take in local populations
between reproduction (baby making, the supply side that keeps
populations going) and economic success (which spills over to
success in baby making, the original and completely valid version of natural selection) is echoed in spectacular ways as you
look at the history of progressively higher-scaled systems in both
the ecological and the genealogical sides of the dual ledger of
life.[1] I'll devote the remainder of this chapter to examples of
pattern no. 3 on all scales—examples that lead us to the "sloshing bucket" vision of the evolutionary process.

## Life's Many Little — and Sometimes Very Large — Vicissitudes

What, then, is this third pattern? It is the near-simultaneous
effects wrought by external environmental events on *different*
genealogical systems represented in a *single* economic system.
That single economic system might be a very small local ecosystem composed of avatars of only a handful of different species,
or it might be the entire biosphere, with all ten million plus
species on Earth at any one time. The effects—and *evolutionary*
effects—are very different according to scale, and a critical

threshold level determines whether or not evolutionary change will be marked and have a shot at permanency.

First things first: take the phenomenon of *ecological succession*. Consider a fire laying bare the hillsides of Colorado, Oregon, Arizona, and all the other western states that have been ablaze in recent years because—ultimately—rainfall has been scarcer than in previous years of recorded history. Though many of these wildfires were started by humans (and not, unfortunately, only through carelessness), lightning is the primordial fire starter and still plays a big role in setting forests and grasslands ablaze. So we have come to think, with good reason, that fire is a natural component in the cycle of life.[2]

All dead wood eventually decomposes, though it takes the concerted effort of fungi and various microbes to get the job done.[3] Sometimes, when the decomposition is *very* slow, plant material accumulates in semidecomposed form, forming peat bogs and layers of coal of varying degrees of compaction. Normal times yield, naturally enough, a sort of steady-state rate of decay, and some forests have more leaf litter, dead branches, and fallen tree trunks than others. Fire simply gets the job done quicker, and the nutrients (if not washed away in the heightened erosion that often ensues) are dumped back into the soil.

And then starts the job of rebuilding—not in neat, human-planted rows as logging companies sometimes (not often enough!) do after a clear-cut operation but by Mother Nature the old-fashioned way: by plants springing up from seeds and spores in the ground that survived the fire,[4] as well as seeds and spores brought in on the wind or in the hair, feathers, and guts of animals that begin to wend their way back into the burned-out remains of their old haunts.

Not all episodes of ecological succession are alike, of course; even if the same patch gets burned back again, there's always

the randomness of what is able to get back there and start growing first that virtually guarantees a somewhat different course to
each episode of succession.[5] But some systems, at least, after an
environmental downgrade like a fire, are started by colonists
that are either really good at moving around (thus stand a good
chance of getting there first) or, even more interestingly, can
establish a foothold better than other species because the chemistry of the degraded environment is right for them or because
they don't need (indeed, do better without) the presence of
other species "normally" there. These are the so-called pioneers,
and it is the fate of many a pioneer species to explode into large
numbers once established—but eventually, inevitably, to
become rare as the later arrivals take up residence, crowding
out their forerunners as the rebuilding ecosystem "matures." In
time, what's there is a forest (or a grassland, coral reef, etc.) that
looks pretty much the way it did before the fire (or hurricane,
etc.) struck—populated by pretty much the same mix of species,
and in pretty much the same relative numbers. Not an exact
replica, but fairly close.

What, in an evolutionary sense, is going on here? My colleagues who study the evolutionary genetics of local populations
are, of course, right when they say that gene frequencies are
constantly in a state of flux in all local populations—and that's
when the environment is in its normal, unperturbed state. So
naturally, when local populations are all of a sudden snuffed out,
or at least sufficiently disturbed that extensive rebuilding kicks
in, the recruits are coming in from neighboring avatars/demes—
and perhaps mixing with individuals that might have managed
to survive. The last two moose in the Adirondack Mountains of
northern New York State were shot in the 1880s, but increasingly moose have been coming back in, from Canada, Vermont,
and elsewhere. Moose genes are being mingled from a variety of

sources as the newly reconstituted population struggles to gain a foothold. Under these sorts of circumstances—with potentially new genes in a new (to those genes) environment—selection is bound to be even more active than under the business-as-usual regimen of long-running ecosystems.

And yet there are reasons not to anticipate any great evolutionary changes during the process of local ecological succession. For one thing, neighboring demes are sending in recruits whose evolutionarily honed adaptations are already suited to the environment they are invading. That's why they are able to get to, settle, and perhaps (re)establish breeding populations. New alleles might well be present, the evolutionary dynamics of gene frequencies might differ quite radically from what they were in the old, now destroyed populations, but in terms of the anatomical features that constitute the visible portion of any organism's economic adaptations, radical departures are not to be anticipated in any great numbers.

But they probably do happen, as we'll see they do in substantial numbers when we ratchet up the intensity of ecological disaster and recovery. There comes a point when "recovery" becomes more evolutionary than purely ecological (via recruitment). But let's stick with local ecosystems just for the moment: Even if some significant, palpable evolutionary change were to pop up during the course of local succession/ecosystem recovery, what would be its fate? If it were to become established as a local evolutionary novelty, it would be a new part of the overall geographic variation within the species. It might spread to other demes, and perhaps even throughout the entire species, though such episodes are difficult to document and are seemingly rare: after all, as we saw in the preceding chapter, local adaptations fit local habitats—and conditions are often quite different from place to place within the entire range of any given species. And, as Darwin himself was probably the first to point out,[6] most variation within species is doomed to be lost. In a glimmer of insight that went far beyond his initial (and stupendously encompassing!) articulation of evolutionary processes and history, Darwin actually saw that new variation is likely to be lost in evolutionary history *unless it is associated with the development of a new species*—a new species that will, in a sense, have its own opportunity to make it or break it independently as the ages roll by. When novelties arise in the course of the development of new species, they are injected into the phylogenetic (evolutionary) stream and have a greater chance of surviving than if left as localized parts of the complex mosaic of geographic variation within a far-flung, well-entrenched species that has already been around for hundreds of thousands, perhaps even millions, of years.

So local succession doesn't produce much, if any, lasting economic adaptive change—at least of the sort we expect to see as detectable changes in the anatomies of animals and plants. What happens when things are intensified—when, instead of local forest fires and tornadoes, or even more regional hurricanes, we consider the effects of, say, protracted climate change?

While Darwin was busy establishing to the thinking world's satisfaction that life had evolved, geologists were busy showing that the history of Earth could also be understood in a rational way.[7] Though most of these early geologists were opposed to notions of biological evolution, they were by no means reluctant to show how much physical change Earth's surface routinely undergoes over time. So Darwin had time *and* environmental change at his disposal, and he painted such a vivid picture of natural selection in the context of geography, time, and environmental change that it was impossible (again, for the thinking world) *not* to agree with him that life must have evolved.

One of Darwin's contemporaries was Louis Agassiz, a Swiss naturalist who established beyond a shadow of doubt that Europe[8] had been intensely glaciated in the geological past. When the reality that enormously thick (up to half a mile) sheets of ice had crept down from the Arctic as far south as present-day New York (Long Island is the composite "terminal moraine" from two of the four most prodigious ice fields) eventually penetrated biologists' minds, it was no real mental leap to see that significant environmental change had been, in a sense, the norm in geological history—and therefore must have had an enormous effect on biological evolutionary history.

And that, of course, is true. But the *way* the connection between physical environmental change and reactive biological evolution was at first imagined has required a significant overhaul. As I already mentioned, Darwin's initial model was simply that (and as we would say today, given sufficient genetic variation on hand) natural selection would "track" environmental change by modifying the features of organisms to keep them "fit" in both the old and the modern senses of the term.

What that dear old original evolutionary model missed, though, was the observation that *what tracks environmental change is the distributions of species—not their anatomical adaptations.* In other words, even when something *major* like a continental ice sheet starts invading a region—mowing down trees, leveling hills, and otherwise transforming a landscape—the whole phenomenon looks for the most part like a gigantic case of ecological succession. As these glacial fields advanced (four major times in the last 1.65 million years), what is now the treeless tundra of northern Canada was splayed out in front of the ice field's margin, which stretched from New York across New Jersey, and on through the Midwest. Below the tundra was the equivalent of the "boreal" (i.e., northern) forest in Canada and a few sections of the continental United States. And so on.

Not that these ecological regions all obediently marched in a row in a nice smooth way as the glaciers inched their way southward—and moved back northward in due course. Far from it. Animals move faster than plants (but plants do move—as they send out their seeds to colonize new areas, much like what happens after fires destroy local ecosystems). So it's a bit more like a patchwork quilt when ecosystems are moving a distance of half a continent. It is hard to say with complete confidence what the ecosystem looked like when the tundra was where New York City now stands (a mere eighteen thousand years ago). It is probably safe to say, though, that the species composition of the southernmost tundra was more different from its modern (or preceding) northern manifestation (even allowing for the extinction in the interim of mammoths and other huge Pleistocene mammals) than a newly rebuilt forest is from its predecessor, thirty years after its destruction by fire.

That said, stability seems to be the norm even in these hemisphere-wide glacial events triggered by episodes of global cooling. Paleontologists are struck by how few species seem to have gone extinct, or were newly evolved, during the Pleistocene.[9] What seems to have happened for the most part—even as the landscape was dramatically transformed—is that organisms were able to keep pace, moving southward and then northward in a dance with the ice. As they moved, they relocated to places that were sufficiently familiar in "feel." They were simply tracking habitats that provided a close enough fit to their already in place economic adaptations, their fundamental needs for making a living, whatever they were. "Habitat tracking" is a sort of messier version of ecological succession writ large, over vaster tracts and more complex environmental mosaics.

So, organisms track habitat movement, vastly changing distribution patterns of their entire species. Natural selection is operating, but still seems to be, at least for the most part, in a "stabilizing" mode. The old model was wrong: under such circumstances, natural selection is *not* doing the tracking. The key to surviving episodes of climate change even as monumental as the Pleistocene glacial advances is to find a place where existing adaptations (familiar prey for a bobcat, suitable soil chemistry for a fern species) can be put into use. It is local populations of species—the demes/avatars—that actually do the habitat tracking.

Unless a threshold is reached: if environmental change goes too far too fast, things become *very* interesting. In addition to the customary habitat tracking in and out as ecosystems are transformed, larger-scale environmental events also trigger spasms of extinction of existing species—and evolution of new ones.

## Critical Thresholds

Take a look at any standard chart of geological time. You'll see a mind-boggling, seemingly limitless series of divisions within divisions within divisions. The grossest divisions since the evolution of complex animal and plant life are the three famous "eras": Paleozoic ("Ancient Life"), Mesozoic ("Middle Life"), and Cenozoic ("Recent Life"). The Paleozoic started with the famous "explosion" of multicellular animals—an extraordinary event of proliferation that took place in perhaps as few as 10 million years.[10] It ended 245 million years ago—in the greatest mass extinction of all time, when at least 70 percent, and perhaps as many as 95 percent (or even more), of the species on Earth at the time were driven to extinction.

Likewise, the Mesozoic (the "Age of Dinosaurs") came to an

abrupt end 65 million years ago, almost certainly as the result of a cataclysmic collision between one or more comets and Earth.

But it is the finer subdivisions of geological time that are particularly arresting. Creationists to this day insist that the standard chart of geological time was developed by people who were interested in "proving" evolution. Nothing could be further from the truth, since most of the early geologists, working long before Darwin published his *Origin* in 1859, were religious (indeed, some, like Darwin's mentor Adam Sedgwick, were clergymen) and, to the extent that their opinions are known, downright anti-evolutionary. In short, they were creationists!

These early geologists realized that fossils overwhelmingly come in layers—"zones"—and that, though some species are found in more than one zone, most species are restricted to a certain segment of the geological rock column. And that is very interesting. The early geologists mapped these zones of fossils and divided up geological time accordingly. They knew they were roughly in the same slice of time when they found fossils of a particular zone as they traveled—around their own bailiwicks, at first (most early work was done in Britain and France), and later as they combed out over the Northern Hemisphere and began finding similar zones with similar fossils in such far-flung places as New York State, Sweden, and China.

So, empirically, the history of life is divided up into packages of time. In the marine realm, at least in the Paleozoic rocks I know best, these time slabs average out to five to seven million years long. For example, in the Middle Paleozoic of eastern and central North America, eight of these packages of life and time are stacked up over a substantial chunk of the mid-Paleozoic. Fossils are abundant in many of them: over three hundred species belonging to many different groups, for example, are known from the Middle Devonian. On the average,

70–85 percent of the species are present throughout each entire interval. The vast majority of them are in stasis (it is astonishing to find clams, snails, brachiopods, trilobites, and other fossils looking very much the same as you pick them up in a quarry that is, say, six million years younger than the lowest level in which you find these same species).[11]

Yet only (again, on the average) 20 percent of the species from each interval survive to make it into the next interval. The paleontologist Elisabeth Vrba, an expert on African antelope and hominid evolution, calls these events—the boundaries between the intervals of stability—"turnovers." Her analysis of a particularly striking example that happened roughly 2.5 million years ago in eastern and southern Africa sheds light on the dynamics of the evolutionary process in general. As an early drop in global temperatures deepened over a period of several hundred thousand years, the African ecosystems continued to hang on—until, rather abruptly, the woodlands of the warm, wet original climate gave way to quickly spreading grasslands as the newer, cooler, drier conditions took hold. Habitat tracking was rampant, as some species left and others came in.

But more than just habitat tracking was going on: the change went too far too fast, and many species apparently became extinct. Moreover, a number of new species seem to have evolved as well: not all new arrivals came in from elsewhere, already adapted to the grassland setting. Some were born there.

Vrba points out that rapid transformation of vast stretches of habitat (and eastern and southern Africa is one vast stretch of land!) would almost inevitably mean that mosaics of older and newer habitats would initially form—ideal conditions for the fragmentation of species. Stands of woodland would linger, now separated as islands in a sea of grass, and serving to fragment populations of mammals—and leading to the evolution of new species. Novel evolutionary adaptations in these small, fledgling species stand a much better chance of surviving than novelties that show up later in larger, well-established species, where they are destined to be swallowed up into the greater mélange of preexisting genetic variation.

Turnovers pepper the entire history of animal and plant life. I am convinced that most of the change we can document in the economic adaptations of organisms—meaning most visible evolution—comes during these rare and rapid turnover events—*events that overwhelm regional ecological systems to the point where entire species disappear*. In fact, it very much appears that most speciation events in the history of life are (1) concentrated in brief spurts in regional ecosystems that (2) involve the extinction of many of the preexisting species of the region.

## Crises in the History of Life

Though turnovers are an empirical reality in the history of life, there's good reason to have predicted that they simply *must* have occurred. On the one hand, we have localized disturbance and the death of local populations—remedied, as usual, by ecological succession. On the other, far grosser scale, we have mass extinctions.[12] It is certain that mammals did not radiate into the spectrum of herbivores, carnivores, scavengers; fliers (bats), swimmers (whales and others), burrowers, runners and lumberers; small, medium, and huge that we see today and that pretty much became the norm, until 5 or 6 million years after the terrestrial dinosaurs breathed their last breath, 65 million years ago. Mammals had first shown up in the Upper Triassic, just when the first dinosaurs evolved. For reasons nobody knows, it was the dinosaurs (and their extended kin) that radiated into all the sizes, feeding types, and so forth. Mammals, including long-extinct forms such as the "multituberculates" and even some primitive eutherians, such as insectivores and even our own group, the primates, remained smallish and undifferentiated. It is blindingly obvious that the prodigious amount of mammalian evolution—including the emergence of our own lineage—would not, could not, have taken place unless the dinosaurs and so many other dominant forms of Mesozoic life had been swept away in a paroxysm of mass extinction.

It is no less obvious that little or no discernible evolution occurs as ecological succession patches up localized environmental destruction. So there *must* be a middle ground where environmental events perturb regions sufficiently large to encompass the entire geographic extent of many species—and happen so intensely that many, perhaps sometimes most, of those species are driven to extinction. There must be something midway between, on the one hand, the ecological recovery, fed by surviving demes, seen in succession, and even in regional disturbances where habitat tracking keeps systems going while the scene shifts around, and, on the other hand, the all-out revolutions of truly global mass extinctions (of which there have been five so far in the history of multicellular plant and animal life, the aforementioned end-Permian and end-Cretaceous being the most famous). Vrba's turnovers *must* have happened repeatedly in the history of life, and we should have predicted their existence even if we had lacked the evidence we have in such abundance that they happen with great regularity as the ages roll.

There is great symmetry here: localized ecological disturbance, no matter how devastating, results in little or no change; ratchet up the process to encompass entire regions, and species start to disappear, to be replaced by newly evolved species. Opening up full throttle, we have our global mass extinctions—

where entire larger groups ("higher taxa") succumb, to be replaced by other higher taxa that, in some sense, at least, can be seen to take their place.[13] The larger the ranks of the higher taxa that become extinct in a mass extinction event, as a general rule of thumb, the higher the rank of the taxa that evolve in their stead—and the proportionately greater the actual amount of evolutionary change that has occurred.

Regional turnovers—the extinction of old, and evolution of new, species—do not as a rule encompass entire genera, families, and so on. The ecosystems that are rebuilt, unlike local ecosystems restored through succession, have newly evolved species among the players in the ecological arena. But successions of turnovers really amount to variant versions of the same basic faunal and floral aspect of the region—an effect enhanced by the 20 percent or so of species that survive to live in two (sometimes more) successive periods.

## THE SLOSHING BUCKET

This is the "sloshing bucket" view of evolutionary history (figure 4)—my explanation of how economic and genealogical processes intersect to produce evolution. No disturbance, little lasting evolution; local disturbance, little lasting evolution; regional disturbance (but no extinction), some evolution, but not much; regional extinctions of entire species, rapid proliferation of new species; global mass extinction, loss of entire higher taxa—and the proliferation (after a characteristic lag of several million years) of entirely new higher taxa of roughly the same rank.

The bucket sloshes as the vector triggering all of this comes in from the outside world; depending on where it impacts the ecological hierarchy, it affects local populations, entire species,

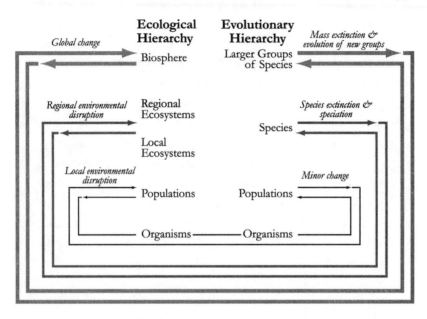

FIGURE 4. The "sloshing bucket" view of evolution mapped onto the twin hierarchies of ecology and evolution. Local disruption of ecosystems (e.g., forest fires) kills off some or all the members of local populations—which are eventually replaced by recruitment from neighboring, undisturbed demes through the process of ecological succession—with little or no conspicuous evolutionary change resulting (*thin lines*). In contrast, global disturbance of ecosystems results in the extinction of entire groups of organisms (e.g., Cretaceous dinosaurs), eventually resulting in the evolution of new groups from survivors of the mass extinction (*thick lines*). Regional turnovers—intermediate in scope and intensity between local ecological disruption and global mass extinctions (e.g., the African turnover of 2.5 million years ago)— are at the threshold level where many species become extinct and new species evolve. Most of life's evolutionary history appears to be concentrated in these turnover events (*intermediate lines*).

or perhaps even entire higher taxa; the greater the disturbance to the genetic diversity ensconced in all those taxa, the greater the change in complexion of the species that evolve in response. Not for nothing did the pre-Darwinian geologists recognize the Paleozoic, Mesozoic, and Cenozoic—for extinction and subsequent evolution conspired to make life look *very* different in those basic divisions—differences that any observer could see.

*Genes don't drive evolution; the environment does.* But genes do supply the raw materials from which life evolves anew. There is a yin and yang here, not a one-way street.

There is one more crucial connection between economics and reproduction that takes us still closer to the question "Why do people have sex?"

# SIX

# Clones, Colonies, and Social Life

Humans are intensely social creatures, and asking why we do anything at all, certainly including having sex, raises several questions. Why are we social? What effect does living in a society have on our basic actions, including sex? And what, for that matter, is a social system anyway?

Lots of animals besides humans are social beings. Bees and ants are social; so are bottle-nosed dolphins, killer whales, and red howler monkeys. All the species of great apes are social, to one degree or another, with the exception of the several species of gibbons of Southeast Asia. Why do some kinds of animals live in groups, while others don't?

It will come as no surprise that sociobiologists see social systems primarily as breeding cooperatives—systems set up fundamentally to aid and abet the spread of genes. Most of the extensive literature on sociobiology is devoted to the reproductive side of the ledger—toward the cooperative behavior seen in troops of baboons, packs of wild dogs, and flocks of birds. Even when the cooperation extends to hunting (as in the already encountered example of the wild dogs that had their cooperatively killed impala snatched away by momentarily cooperating spotted hyenas), the hunting is seen as a necessary component of the real business at hand: the economics of the group is all about the reproductive success of the members of the troop—the genuine bottom line according to the fundamental theorem of selfish-gene biology.

Ah, but even the more enthusiastic would-be appliers of selfish genes to understanding cooperative behavior in the animal world were stymied from the get-go by one annoying conceptual roadblock. To get to the point where social behavior could be interpreted essentially as a gene-spreading mechanism, evolutionary biologists had to confront and solve a major paradox that had bedeviled evolutionary thinking since its inception. Why, if life is a struggle for existence, if competition for resources and mates is what the game of life is all about, is there any behavior that could look like cooperation at all? How can cooperation do an organism any good when the name of the game—ever since Darwin first articulated the concept of natural selection—is competition between organisms for resources and, ultimately, reproductive success? Darwin himself had openly wondered why there are so many signs of cooperation, bordering on self-sacrificing "altruism" in nature. What's in it for the individual, still more, for the selfish gene, to cooperate?

Good question. In the 1960s, as it became fashionable to see evolution as the outcome of genes shoving one another aside in the race to make it to the next generation, the paradox of altruism finally seemed to have been solved. Animals should be selflessly cooperative in direct proportion to the percentage of genes they share.[1]

The idea makes good intuitive sense and to a degree fits the situation in nature. E. O. Wilson led the charge to apply this hard-core, gene-centered view of evolution to the world of social systems; he coined the term "sociobiology" in his epochal, eponymous book *Sociobiology* (1975). Wilson is an expert on ants, so it was natural that they, plus their close hymenopteran bee and wasp relatives, were the primary focus of his initial efforts to explain the structure, function, and evolution of social systems in general.

How accurate a picture of the natural world is it to claim that cooperation tends to vary in proportion to the degree of genetic similarity? The answer depends on the system you are looking at.

## Bring On the Clones

Consider, first, a clone of bacteria, such as the common intestinal denizen *Escherichia coli*. Whether in the wild inside in your gut or cultured in a petri dish in a laboratory, bacteria such as *E. coli* live their lives as single, separate organisms. And while there's a mixture of genetic types of *E. coli* in your gut (including the odd stray from tainted hamburger meat that, instead of helping you digest, may actually kill you), petri dish experimental cultures are usually pure strains—which is to say, all the individual bacteria are clonal copies of each other. The only genetic variation comes in through mutation as the clonal colony is kept alive through many generations.[2] Bacteria can exchange genetic information, but reproduction for the most part proceeds by simple division, where one simply splits into two genetically identical individuals. Each bacterium spends its life absorbing and metabolizing whatever nutrient and energy source is available (usually rigidly controlled under experimental conditions—and much more wildly and widely variable in a human gut); bacteria don't grow except right after their "birth" as a two-for-one split, so it might appear that the whole purpose of a bacterium's existence is to make more bacteria. But no need to invoke the fundamental theorem of the selfish gene too quickly here, since it is (once again) just as easy to maintain that the entire "purpose" of a bacterium's life is to get and metabolize whatever nutrient and energy sources it can find—reproduction coming only as a luxury afforded to the bacteria that succeed in their economic mission.

Bacteria are often said to be immortal—because they keep on dividing, and dividing, and dividing—and no one has yet identified anything remotely akin to the mechanisms of cell aging and death that are built into the eukaryotic cells of multicellular organisms. Theoretically, it is possible that the actual physical ingredients of a single bacterium—its single-stranded DNA chain, its cellular wall, and the rest of its simple internal anatomy—could be passed down over the millennia and through long periods of geological time. Theoretically possible—but highly unlikely, given the volatility of life in the wild, and even under carefully controlled experimental conditions. What is potentially immortal, as we have already seen, is the information carried by the genes.

Then, too, bacteria are not famous for cooperation. Despite the genetic identity uniting all members of a clone, there is not a scintilla of evidence that bacteria in a petri dish are doing anything more than each going it alone, gobbling up the supplied nutrients until they are exhausted. One favorite kind of experiment has always been to introduce a medium (energy/nutrient source) known to be either downright poisonous to, or at least unutilizable by, the strain (genetic variety) of bacterium being cultured. If a mutation has occurred in the clonal culture's history that perchance enables a few of the bacteria to metabolize the new medium, then they will "take off" and the other ones, adapted to the original medium, will quickly die off—a simple, graphic demonstration in the laboratory of natural selection in action.

So genetic similarity, up to and including truly clonal genetic identity, is not automatically associated with cooperation; nor does it ineluctably, automatically lead to anything resembling altruistic behavior. Blue-green "algae" (bacteria, actually) do form mats of attached individuals. These are among the oldest organisms found in the fossil record, and certainly the oldest that a paleontologist can find sticking boldly out of a rocky Precambrian outcrop. Such "stromatolites" are still being formed in some shallow tropical marine environments today, such as the famous ones of Shark's Bay in Australia. Every morning, these photosynthesizing bacteria start madly metabolizing and dividing (reproducing by splitting), building up an organic filmy layer that, when the sun sets and the algae perforce take the night off from their labors, is often covered by a layer of sediment washed over them by the sea. The next day, they are back at it, and another layer is produced on top of the old—eventually building increments several feet thick, all built on thin, daily layers. There is indeed a sort of prototypical cooperation here, in that the bacteria are communally creating the working environment in which they can figure to revive the next morning and get on with living their semiconnected lives.

## Next: The Eukaryotes

With eukaryotic cellular life, things get a whole lot more complicated. The earliest forms of eukaryotic life—still very

much with us today, and still very numerically predominant—
are single-celled, usually microscopic (though far larger than
bacteria) protoctists such as diatoms, formanifera, amoebae, cil-
iates, and flagellates. Most of these are free living in the wild,
though some cause nasty familiar diseases like malaria. Some
photosynthesize and are quite plantlike (the flagellate *Euglena*,
for example), while others (like amoebae) engulf bacteria and
other protoctists and thus behave, economically, more like animals. Still others can do both.

Difficult as it is to study the reproductive behaviors of the
protoctists (especially, perhaps, the marine plankton that have
proven hard to culture in the lab), it has become fairly clear
that protoctists often can have it both ways: though simple,
asexual fissioning is common, and perhaps even the "usual"
means of reproduction of any one species, it is dangerous to
assume that protoctists invariably form only genetically identical clones. They can—and at least semiregularly do—exchange
genetic information between individuals *before* they split.

And there are, as well, famous examples of simple protoctistan colonies, such as the beautiful pond-dwelling *Volvox*, gently
rotating balls of life formed by genetically identical individual
cells of a photosynthesizing flagellate protoctistan. Though no
nerves unite the cells to coordinate their activities, some form
of integrative communication is present, as the flagellae from
each cell, which collectively propel each *Volvox* colony through
the water, are coordinated to propel the colony in any one particular direction.

## Next: The Multicellular Body

We are building a picture of increasing anatomical complexity and of increasing cooperation among cells. All this coopera-

tion is fundamentally *economic* in nature: there is as yet little sign among these economically cooperating, generally genetically identical cells that they are also cooperating to make additional sets of cooperating cells. They are simply dividing as individuals and otherwise forming a communal, economically cooperative architectural structure.

That is, until we reach the phase of true multicellularity—in plants, fungi, and animals. Think of it: your body is one giant mass of clonal cells—billions of them, differentiated into hundreds of particular types, each to perform a specific, restricted set of economic functions. A liver cell is busy knocking out excessive alcohol by making alcohol dehydrogenase, while a red blood cell (the only type of cell in your body that actually has dispensed with nuclear genes, meaning that red blood cells themselves don't reproduce) is bringing oxygen to your tissues so that basic metabolism can continue in every one of them. Meanwhile, nerve cells in your brain are busy firing off signals to integrate your basic bodily functions *and* allow you to see the car bearing down on you as you step off the curb into traffic.

Why does your liver cooperate with the rest of your body to regulate the amount of sugar and other substances in your bloodstream? Animal bodies are economic machines, and the more complicated ones perform all those economic animal acts by assigning specific tasks to particular organ systems (the digestive system, for example), organs, tissues, and cells.[3] Such an intricate division of labor—for that's what an animal's body represents—can evolve only, one might credibly argue, if the genetic content of all those cells is identical. Cells are derived from preexisting cells through the same basic process of cell division used by amoebae, and the economic cooperation seen in simple bacterial blue-green "algal" mats and, in somewhat more complicated form, in *Volvox* colonies has by now become many times more complex.

These myriad cells, with all their different anatomies and internal chemistries, are derived from a single sperm-fertilized egg cell—the zygote. And that remains perhaps still the greatest, amazing, and as yet incompletely solved mystery of all biology: How can a single fertilized egg, with matching pairs of chromosomes, one set from the mother, the other from the father, develop into such a complex, yet finely integrated, system of two hundred different kinds of cells, each performing a specialized function as constituents of different sorts of tissues and organs? There's no problem multiplying through cell division, but we are not gigantic, billion-celled fertilized eggs. How can cells differentiate into different types if they all have exactly the same genetic makeup?

That's one of life's mysteries I cannot solve here, though molecular and developmental biologists have been making enormous strides in understanding how the same genetic information can be expressed, in a controlled, regular way, to produce a frog, a leopard, or a human being from a single fertilized egg. Suffice it to say that each of us—each individual frog, leopard, or human being—is a gigantic clonal colony, each with a body amazingly differentiated to perform subsets of the economic tasks that fall to every organism simply to exist: to be born, develop, and stay alive.

And now also to reproduce. Here is the strongest claim that can be put forward in support of the general approach to altruism advanced by evolutionary biologists since the 1960s. It is a claim, however, not about evolution but about the material fact of existence of any multicellular organism. All of an individual's cells (mostly economic, somatic—but also including the reproductive, germ-line half-genetically complemented sperm and egg cells), being genetically identical, act in concerted interest for the maintenance of life of the collective whole—the individual organism. The big exceptions to genetic identity of all cells are the intervening mutations; and the huge exceptions to the theme of universal cooperation among your body's cells are those cells leading to cancer—cells with their own agenda, so to speak, cells that are out to destroy the body. But when it comes to the normal, differentiated cells of a multicellular organism's body, it definitely is all for one and one for all, no question about that.

And that goes, as well, for reproduction: Weismann's germ-line cells, the sperm and eggs. They are carrying, albeit at half complement each, the same genes that make up your liver cells, hair cells, heart muscle cells. Their specific role, however, is not economic; it is reproductive. And although it would be tempting to score one here for the notion that organisms are especially "concerned" to spread as many copies of their genes as they can to the next generation, it should at least be asked, "What other genes could an organism spread?" What if the built-in propensity were merely to reproduce—and to do it as well as possible, given whatever handicaps (whether internal deficiencies, simple bad luck, or environmental hardships) may afflict any particular single organism in its urge to reproduce? Indeed, it is again easy to see the advantage to an organism if its nonreproductive, somatic economic cells are cooperating in harmony. That means the body is managing to stay alive—perhaps even thriving.

But no one in evolutionary biology, so far as I can tell, has provided a cogent answer to this question: Of what benefit, material or otherwise, is it to an organism if its genes survive to the next generation, rather than, say, merely being spent on the ground in failed mating attempts?[4]

## STICKING TOGETHER:
## THE FIRST CLONAL COLONIALISTS

The human body, marvelously complexly differentiated clonal structure that it is, is still not the epitome of clonal cooperation. That distinction goes to certain forms of marine invertebrate colonial organisms. These colonies grow by asexual budding of new individuals from parental polyps or zoids, depending on what kind of animal we're talking about. So, like the cells within their own bodies, all multicellular individuals in a colony are also genetically identical. But the trick is that some of these colonies show a remarkable degree of anatomical differentiation and economic specialization. Take jellyfish, for example—the floating relatives of corals and sea anemones. Though some are simple, single individuals, many of the larger ones, such as the infamous man-of-war, or the lion's mane, not only are large but are actually colonies, with some individuals (often equipped with their infamous stinging cells) used for immobilizing prey and colony protection, while others are devoted to digesting the booty of hunting, or for reproduction. The whole thing looks like a single jellyfish, and in a sense, of course, it is. Yet in another sense these complex colonial "super"-organisms are even more complicated than a human body, consisting, as each one of them does, of a number of genetically identical, yet highly differentiated, individuals, each one of which is itself a multicelled, differentiated animal body in its own right.

Corals and jellyfish have no truly differentiated organ systems—that is, for digestion, excretion, and so on (though a simple nerve net does integrate the motions of an individual polyp). Bryozoans are far more complex, with organs for feeding (the "lophophore"), excretion, digestion, and the like. Bryozoans are simpler versions of typically differentiated multicellular animals. Yet all bryozoans live in colonies of genetically identical zoids, which are often differentiated to a very high degree. Some— the majority—are for feeding; others, looking very different from the feeders, clean and guard the colony, while others indulge in the sexual phase of reproductory matters.

These colonies are in a sense "superorganisms"—the colony itself is like a single, differentiated metazoan individual—yet its component parts are not organ systems, but entire individuals. The mind fairly boggles at this hierarchy of complexity and division of labor—all done by clonal, albeit differentiated, individual organisms!

## Separately Together: Insect Colonies

Let us now turn to the primordial inspiration for Wilson's sociobiology—his much beloved ant, bee, and wasp hymenopteran kin. Social ants and bees are genetically unusual. The females, the workers of various castes, are more closely related to one another and to their queen mother than to their father or brothers because they share more genes. That's why, says sociobiology, all these sterile female workers will do so much work for the colony, sacrificing their own reproductive potential because *that* job is being handled nicely by the queen and her consorts. The worker females can go out and get nectar, keep the hive clean, and tend the developing larvae, all because they share so many of their genes with the queen's new babies. And there is something to this line of thought.

One of my favorite papers in the annals of sociobiological

research is Wilson's comparative analysis of complexity—division of labor—among worker females among a series of insect colonies of varying complexity.[5] Some species have only a few "castes" of workers—while labor is divided (foraging for nectar, tending the eggs, cleaning the nest) far more minutely in the more advanced, complex hives. In some species, the workers actually shuttle their way successfully through all the castes as they go through their seasonal lifetimes. Fascinating! And all about economics, of course—or at least at first glance. For if we grant that the peculiar form of genetic inheritance among female hymenopterans makes it feasible to leave the egg laying to one superfemale (the queen), freeing the other females to various aspects of economic housekeeping, the examples also reveal one ineluctable truth about social systems generally—a truth often glossed over in most sociobiological studies: *social systems among nonclonal animals are fusions of the economic and reproductive interests of their component individuals*. This is especially true where degrees of genetic relatedness form a major, if not the sole (i.e., as in clonal situations, like your body, or a bryozoan colony), determinant to the actual structure and function of the colony itself.

In other words, social groups are fusions of economic and reproductive life, mergers of the eating and baby-making adaptations of species (as yet another version of the dual hierarchy scheme shows in figure 5). Small, localized groups that they are, societies—be they beehives, killer whale pods, or monkey troops—are hybrid populations, acting at once as economic units (avatars) within the local ecosystem and as breeding cooperatives (demes) within a species. Here is a connection par excellence between the economic and the evolutionary hierarchies: each local beehive is playing a concerted role in the local economy of nature—the exchange of matter and energy among

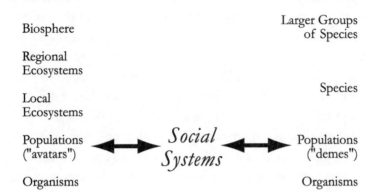

**FIGURE 5.** Social systems are population-level groups that intermingle and integrate the ecological properties of avatars and the reproductive properties of demes.

populations of different species within the local ecosystem.

So it makes no sense to think (as sociobiologists routinely do) of animal societies strictly as breeding cooperatives. No more sense than it does to think of humans purely as sex machines. Some people some of the time may be so immersed in their sex lives that they seem like pure sex machines, but even Casanova had to eat, and humans are really no different from lions or armadillos. Though sex plays a greater role in the average human life than in the lives of nearly all other creatures, we humans are right in there with the rest of life in that our lives are much more about economics than about sex, whether sex is about reproduction or something else. Indeed, it is because so much of human sexuality is about economics, about making a living, that we spend much more time than lions or armadillos do in all manner of sexual indulgences.

So, attractive as unusual insect genetics may be to explain the apparent "altruism" embodied in female worker ant and bee economic behavior, for the welfare of the group as a whole, it is even a mistake to think of their colonies and hives purely as reproductive cooperatives. A beehive is a lot more than that. Each is in fact a highly condensed and localized population that plays a distinct role in the economy of nature around it. Each hive is a part of, and plays an economic role in, the local ecosystem.[6] Bees are out in the ambient world, collecting nectar, inadvertently pollinating the plants they visit, and being eaten by birds and other predators. Bears, humans, badgers, and other species routinely raid their hives for honey. To claim that bee life outside the hive, and even for the most part within the hive, is purely about reproduction is simply to jump to that same mistaken conclusion: that all economic activity is *really* just about making more babies. Beehives are no different from any other form of life, including social life. Hives are about especially tight integration of the economic and reproductive lives of the organisms that live within them.

## FREE AT LAST? THE EDGY COOPERATION OF VERTEBRATE SOCIETIES

We are edging off that peak of ultimate cooperation based on genetic identity seen in our own bodies, in jellyfish, and in bryozoan colonies. Bees and ants are not clonal, though the vestige of degrees of genetic similarity among the individuals clearly has a lot to do with the way the colony is organized and functions: the roles played by the males and (various sorts of) females. But sociobiology has by no means confined its gaze to situations where a spectrum of genetic relatedness—a measurable spectrum of gene sharing—is playing an obvious role in what is going on. When we turn to vertebrates, much of the easy genes-in-common argument simply no longer obtains, and the "I'll cooperate with you, even sacrifice my own chances to reproduce, because I reproduce vicariously when *you* reproduce" approach doesn't work at all well when it comes to fishes, birds, mammals, including primate mammals—and especially not when it comes to us.

Once more there's the exception that proves the rule. There may be "leks"—where males strut their stuff in front of a large gaggle of female grouse, competing in good sexual-selection form to mate and make babies.[7] And there well may be rigid power hierarchies (based on economic stamina and—who knows?—accumulated wisdom) where only a few dominant bulls get to service the herd.[8] But, with a single exception, nothing in all vertebrate experience remotely resembles the genetics of bee colonies. For, no matter how highly socialized, all mammals and birds mate and make babies just like their nonsocial cousins that get together with other loner adults only once a year, when they hook up to try their hand, as it were, at making babies.

Here's that exception: naked mole rats, tunneling rodents of one or two different species that live in an amazing maze of tunnels somehow excavated from the tough hardpan of the drier regions of eastern and southern Africa. These critters really do live like bees in a hive, with one enormous female acting as sole reproducer of her sex, a baby-making machine fed by her sisters. She dies, and someone else transforms and takes her place. But naked mole rats are the *only* vertebrates whose social system and reproductive habits at all resemble those of social insects, so they are the exception that starkly highlights how different all the rest of the birds and mammals are.

Consider Florida scrub jays. When first studied in the 1970s, they were presented by many observers (though not, ironically, by the authors of the definitive study on Florida scrub jay social biology)[9] to be a poster example of Hamiltonian altruism, cooperation based on degrees of genetic relatedness. Florida scrub jays exhibit "helper" behavior, in which the past one or two broods often hang around to help feed the next batch of kids. Clearly, it has been routinely assumed, they are helping to pass along their own genes, albeit vicariously and not all that efficiently, when helping mom and dad feed their later siblings and defend the territory against invaders.

Ah, but helper behavior, it turns out, is really all about territory, and territory means where you live and eat as much as (even more than, actually) where you mate and raise kids. There is not a lot of territory suitable for Florida scrub jays, so good serviceable spots are rare and hard to come by. Because there is often no readily available spot for the young to go to, they are better off hanging around the old territory, which they frequently "inherit" when the old folks drop out of the scene. It's the territory, stupid, a place to make a living—and if inheriting the territory also yields a better chance of raising your own kids someday, so much the better. Meanwhile, twenty-four hours a day, seven days a week, 365 days a year, you gotta eat, and once again we see that social life is by no means all about making babies.

Much closer to home, in an evolutionary sense, consider the case of *Pan paniscus*, the bonobos, or "pygmy chimps." Smaller than their better-known chimp cousins (*Pan troglodytes*), bonobos live in troops of up to two hundred individuals, an unusually large group size for apes. They are probably our closest living relatives. Though we also share near genetic identity (98.6 percent!) with *troglodytes*, bonobos look even more human than those other chimps. They never develop the large brow ridges and other features of typical adult chimps, retaining instead a babyish quality to their faces. Just like human beings, bonobos are distinctly "neotenic," meaning that we both keep many of the features of babyhood all the way through adulthood and that we look a lot more like our ancestors' children than like our ancestors as adults.

But that's not all we share with bonobos. Bonobos really like sex. They engage in it a lot, often with members of the same sex. They like to masturbate, and to help others do the same, when they are not actually humping one another. A lot of bonobo social life seems to revolve around this near-incessant sexual behavior, and clearly not all, or even much, of it is about making babies per se.

But much of the interaction, including overtly sexual interaction, among bonobo individuals has to do with their foraging behavior, with their daily, constant work to find something to eat. According to those who know them well,[10] bonobos shifted their diets from strictly fruit to include a greater variety of herbs when they split off from other chimps, and thereby reduced their competitive behavior while foraging for food. Males fight a lot less in bonobo society than in other great-ape species, and this economic neutrality (if not downright cooperation) leads to greater contact within larger group sizes. And familiarity breeds, it might be argued, not contempt, but sex. It could easily, though, work the other way too. Friendlier, sexier relations may well be helping to cement the bonobo social contract, our first glimmer of a "use" for sex beyond the pure production of new babies. Unsurprisingly, then, bonobos, the sexiest of social animals apart from ourselves, are our closest relatives.

Social systems are thus economic and reproductive cooperatives. Sometimes, as in hymenopteran colonies, the reproductive side, at least at first glance, is the more obvious. In the vertebrates, the distinctions are less clear-cut, and we really are looking at functional fusions of the otherwise more rigid and distinctive difference between avatars (in local ecosystems) and demes (within species). Cooperation begins with genetic identity as the lives of individuals are commingled into colonies—and into the complexly diversified cells of multicelled plants, fungi, and animals. The cooperation at that level is purely about economics—division of labor to perform subsets of the vital functions faced and performed by all life. Indeed, the deepest forms of cooperation in evolutionary history—the permanent fusion of previously separate forms of life to form single entities—is also profoundly economic in its primary aspect.[11]

With the vertebrates, the cooperation that is a hallmark of social behavior is easily seen as a resonance between economic and reproductive interests. With people, it's even more fun—for now we find that sex has become decoupled in many ways from reproduction. We'll address this human triangle of sex, economics, and reproduction right after we consider how well human biology fits the duality-of-life picture developed so far.

*Part Two*

# Human
# Singularities

# SEVEN

# Naught So Queere as Folke

## *The Strange Biology of*
## *Modern Humans*

How well do humans fit into this neat dichotomy, this duality of ecology and reproduction, which seems to describe the conditions of life for all other organisms on the planet? Like all other animals, we continue to breathe, to seek energy sources (i.e., to eat), and in general to rely on the basic physiological processes absolutely essential for each and every organism simply to stay alive. And, of course, we reproduce—in ever-startling numbers: our population has more than doubled in the past sixty years. There are now more than six billion of us.

Yet we don't play by some of the most fundamental rules in either the ecological or the reproductive/genetic side of life. With few exceptions, we no longer live inside local ecosystems. And our species, unique among the ten million-odd species on Earth, has become an integrated, economic entity.

These peculiarities—for such they certainly are—about the human condition are part of the story of why people have sex.

## CULTURE, LANGUAGE, AND
## THE STRANGE STATE OF HUMAN ECOLOGY

Humans are ecological generalists. We come from a long line of species that eat just about everything. But there is something distinct about the way we have stayed ecologically flexible. Paradoxically, we have done so with an almost extreme anatomical and, consequently, behavioral specialization. The frontal lobes (the "neocortex") of our brain have expanded, and with that expansion has come—in ways yet to be figured out—consciousness. As far as anyone knows, we are the only species in the history of life to be (oh so acutely!) aware of our own impending death. A stark way of defining consciousness, perhaps, but a telling one.

We are not a social species because of consciousness, for we also come from a long line of preconscious (or at least not so thoroughly) conscious social species. But self-awareness—the ability to consult one's inner thoughts to divine what may be going on within the minds of others around us—has become a crucial ingredient of human life. With it has come language: our way of thinking thoughts and communicating them with others. Some exciting new genetic data has linked a particular gene (the "FOXP2" gene) with the ability to speak. The FOXP2 gene has remained very stable through mammalian history (meaning that it is very similar in all modern mammal groups in which it has been studied), but is different in two places in the human version. Apparently these two mutations make all the difference, because humans with defects in these genic segments are incapable of speech. The human version of FOXP2 occurs through the six billion strong modern human population, suggesting to some geneticists that its importance has caused it to spread specieswide through natural selection. Given the calculated age of 100,000 years of the human version of FOXP2, though, it is much more likely that it originated when our species was still confined to Africa—and simply spread in the great human diaspora around the world that began about 100,000 years ago.

Language facilitates and greatly expands the range of cultural items we transmit to one another, and from generation to generation: at first, perhaps, the ways of making and using tools—dimly foreshadowed in our ape kin, and part of the human evolutionary record since the oldest tools appeared 2.5 million years ago at Olduvai Gorge and other locales. Language allowed a much greater range of cultural items and nuances of meaning in their transmission. And it is culture—that highly specialized behavior that has developed from our consciousness, and thus from our expanded thinking brains—that has made us even more ecologically generalized than ever our ancestors were before us.

Culture enabled our ancestor *Homo erectus* to travel north from the tropics in the very teeth of a major glaciation event one million years ago.[1] Fire, clothing, and spears were all that it took to create the greatest expansion of an ecological niche recorded in the entire history of life: species are rarely seen moving into entirely different climatic regimes—let alone moving from the tropics into the frozen steppes of the higher latitudes. And it was culture, learned behavior in the form of the agricultural revolution, that ten thousand years ago took us out and away from the very local ecosystems in which our ancestors and all other species, like spotted hyenas and wild dogs, have always lived.

This is worth pausing over: absolutely every other kind of

organism, from bacteria to sexually reproducing species of plants and animals, lives in groups inside of local ecosystems. But now, all of a sudden, humans have started to do something very different; with the invention of agriculture in a number of different places starting about ten thousand years ago, humans abruptly began to live *outside* local ecosystems. Think of it: to till a field, getting rid of all the native trees, shrubs, and grasses that grow there (not to mention many of the animals that depend on them for food and shelter), and to plant instead one or two species of desirable food plants (einkorn wheat was an early development in the Middle East) is in effect to declare war on the local ecosystem. All plants save those deliberately grown there for food production are now by definition "weeds."

Domestication began some thousands of years earlier—that is, the practice of taming and bending animal and then, a bit later, plant species to our will (the contrary scenario, encountered in chapter 1, that animals deliberately adopted us as a way of surviving the rigors of the Ice Age notwithstanding). With food production, especially planted crops, now underway, humans began to lead a more settled existence, and, with that, to adopt a division of labor. At this first agricultural revolution, the average of the estimates of how many people were on Earth ten thousand years ago is about five to six million. Becoming self-sufficient (and despite the specter of famine that has stalked the human agricultural enterprise since its inception), our population has exploded from five to six million to six billion at the latest millennium.

## Our Evolutionary Psyche

That we have changed the ecological rules in such a dramatic fashion has deep implications for the working of natural selection in modern human populations.[2] In all other species, natural selection and sexual selection operate in local populations firmly within the context of local ecosystems. But if we aren't ourselves living in such systems, what can we say about natural selection acting on *Homo sapiens* in the modern world?

Evolutionary psychologists often speculate that the real evolutionary crucible of *Homo sapiens* behavior lay out on the African savanna during the latter stages of the Ice Age (the Pleistocene epoch). That's why much of our behavior seems so inappropriately out of synch: it was evolved expressly for the purpose of life on the African savannas, but now willy-nilly we are stuck with genetically emblazoned behaviors never evolved for life in the urban jungle. The most famous example—especially apt, for it often (though not always) involves the direct spread of genes or at least the forced intrusion of seminal fluids into vaginal tracts—is rape. Rape, everyone agrees, is inappropriate behavior in the modern world; but it is (according to some evolutionary psychologists) a vestige of human life in the dim recesses of our evolutionary history—when it was a hangover from promiscuous times among our even remoter human ancestors.

We'll get back to rape. For the moment, we want to know how our changed ecological status affects how natural selection is working on us in the here-and-now of the modern world. At first, the "things were evolved in the Ice Age and no longer fit" cry of the evolutionary psychologists looks rather good: for it is certainly true that agriculture, the very basis of modern civilization, has removed us from the confines of the local ecosystem and has resulted, as well, in our living in a world very largely of our own construction.

But what happens when evolved traits become—because circumstances change—dysfunctional, or at least no longer the optimal way? They are rapidly modified or lost altogether, that's what. Though culture has drastically modified our relationship to the world, there are still plenty of circumstances where we see natural selection at work in human populations, as in the famous case of sickle-cell anemia.[3] Allegedly evolved, genetically based traits of human behavior strike me as just as open to removal or modification by natural selection in the modern world as in the days when they were evolved. And anyone who supposes that ten thousand years is too little time for evolution to occur should consider the startling fact that twelve thousand years ago, Lake Victoria dried up almost completely, killing off nearly all the species of fish and other animals and plants that were living in there. For one group alone—the famous cichlid fishes—there are over three hundred distinct and separate species currently living (though now highly endangered) in Lake Victoria alone, all but a handful of them evolved in the last ten to twelve thousand years. Evolution can be very fast indeed.

I think evolutionary psychologists are trying to have it both ways: we see evolutionary traits, however dysfunctional, and we say, "This was evolved in the Ice Age, when this behavior had a material effect on reproductive success—and was selected for." But now we say this trait either has no—or a negative—effect on reproductive success. It's just there, a vestigial holdover from times when it usefully aided the spread of genes. This is the great Pleistocene cop-out of evolutionary psychology: it says that if we analyze human reproductive behavior and its relation with human economics and (the often separate) arena of human sex, we will not uncover the forces that led to the evolution of these behaviors in the first place, because the conditions under which they evolved (supposedly—yet another assumption) no longer obtain. As we will see in abundant detail, the decoupling of sex from reproduction in human beings and the myriad patterns of interplay between sex and reproduction, sex and economics, and reproduction and economics (the "human triangle" of the latter section of this book) at the very least establish without a doubt that much of human behavior has very little to do with baby making—hence the spreading of our genes. And to allege that something might be so now, but wasn't in the past, is at best an interesting speculation (and, as we shall see, a testable one to some extent) but by no means a foregone conclusion.

## EUGENICS, GENETIC ENGINEERING, AND THE HUMAN EVOLUTIONARY FUTURE

Worries about how modern life seems to remove humans from the normal reach of natural selection have been around for a long time. The eugenics movement, largely founded by one of Darwin's nephews, Francis Galton, was concerned that human cultural advances—in nutrition and especially, perhaps, in medicine—meant that the "unfit" were increasingly allowed to live longer and, of course, to breed. Anyone who has worn eyeglasses from an early age (as I have) can only be grateful for such advances, and though poor eyesight might have lowered my fitness, not only in the hunting/gathering cultures of far-away Pleistocene Africa, but perhaps as well in the later farming communities of the Neolithic, the anthropological literature is full of examples of similar, seemingly less fit individuals who were nonetheless functional members of ancient societies—their infirmities in a sense carried by a caring native band.[4] The long and short of it is that culture—learned behavior—does override the pure winnowing features of natural selection. If that means that deleterious genes that might once have been kept to lower frequencies by natural selection have long been allowed to persist in human society in higher frequencies because of cultural means to mitigate their effects, the largely unacknowledged, but implicit, view of humanity at large has been: So what? Culture, in terms of these infirmities, has made them largely, if not wholly, irrelevant. They can even be, in this new context, distinctly useful, as when one's spouse turns off her hearing aid when she knows she doesn't want to hear something.

## FUTURISTIC EVOLUTIONARY FANTASIES

Then there are the mythic projections of what human evolution, if left to go on its own, will end up producing. Ever wonder where the famous movie figure E.T. came from? E.T. is the literal embodiment of the most common of futuristic human evolutionary scenarios in place virtually ever since Darwin convinced the thinking world that life—including humans—has evolved. Big head to house a greatly expanded brain; emaciated, shrunken body, because machines now do all the heavy lifting. Once again, not an unreasonable scenario, as we are, in increasing relative numbers within our logarithmically expanding populations, relying more and more on mental prowess over physical labor in our daily rounds of making a living and staying alive. Indeed, for many of us, sex is probably the most strenuous activity in which we routinely indulge![5]

Will selection end up making our brains bigger and our bodies smaller? Not much chance, because our brains are already so big at birth that the skull bones of infants awaiting birth remain unknitted so that the skull can for a short while be distended to allow passage through the already too narrow pelvic passageway of the female human birth canal. And though it is conceivable that the female pelvis could be enlarged to allow the birth of babies with even larger brains, it hasn't happened in the last

100,000–200,000 years of the existence of *Homo sapiens*. To do so would presumably compromise a woman's ability to get about bipedally—an earlier and still vital aspect of human anatomy, physiology, and behavior. Chinese women are said to be in the rice fields often only hours after giving birth.

We've seen that, anatomically speaking, species that tend to survive for hundreds of thousands—and often millions—of years tend to do so without accruing much, if any, significant anatomical change. The problem for evolutionary change within a species grows with the increasing size of the population—and humans, currently six billion strong, are one enormous population. Our sheer numbers are a powerful impediment to accruing any specieswide anatomical change. If we are to survive, we'll do so looking pretty much as we are today, and have been for over 100,000 years.[6]

So there it is: selection still can—and will—remove deleterious, dysfunctional features and behaviors, except those (like poor eyesight) that are in a sense "protected" by cultural remedies, in effect accepted by human society. But as far as substantial genetic modification of anatomical or behavioral features of the entire human species is concerned, forget about it: the evolutionary decks are too stacked against that.

These thoughts arise from the strange state that human ecology has reached through more than 2.5 million years of (distinctly nongenetic) culture impinging on our behavior, culminating in the agricultural revolution that made us the first species ever to live outside the confines of local ecosystems. The astronomical rise in our population is a simple reflection of that state, as we are no longer limited by the productivity—the abundance of appropriate foods—in local ecosystems.

And now, thanks to our global ubiquity and the information revolution, our ecology has recently adjusted still further. We exchange in excess of one trillion dollars of goods and services among ourselves globally *every day*. This makes us the first species in the 3.5-billion-year history of life on Earth to be integrated economically, as well as genetically. There is more than mere metaphor to the "global village." We remain a part of the nature from which we sprang, though we are now collectively a part of the entire biosphere, rather than being broken up into small populations as parts of strictly local ecosystems. And that implies that evolutionary dynamics—however muted and even modified by culture—still play a role in our daily lives and in the future of our species.

So, ecologically speaking, we are truly a strange manner of beast. Because evolutionary psychology is sociobiology applied expressly to humans, and because sociobiology springs from selfish-gene theory, we are dealing with a collective paradigm invented for the rest of the natural world, and now applied to humans—without, for the most part, taking these huge ecological changes into account.

And that is why the ecological hierarchy continues to be relevant to our story. Our species remains an interactive part not of local ecosystems (which, given our emancipation from, we seem almost gleefully bent on destroying) but of the entire biosphere. We are still part of the ecological hierarchy of life— but our species has a unique relation to the realm of matter–energy flow of the word's ecosystems.

And that is why our descendants down the evolutionary road have little chance of ending up looking like E.T. Stability is the norm for virtually all successful species. Far-flung species might develop quite a bit of diversity, but without speciation to chop that diversity up, there is little likelihood it will accumulate to produce something very new-looking under the evolutionary sun. And it is unlikely that our species will be subdivided into separate species. Far from it: the global economic integration of *Homo sapiens* (with all its glitches and horrible shock waves—like that of 9/11—faults and attendant troubles) is turning us into, if not one big happy family, at least a mega-population where the potential for genetic mixture is far greater than it probably has ever been in the 150,000-odd-year duration of our species so far. Diversity is bound to drop—certainly culturally, and quite possibly genetically as well.

But then there's the fantasy of taking charge of our own evolutionary future through genetic engineering. And there is no doubt that bioengineers have already transformed the genetics, thus the features, of organisms of many species. But that really isn't "evolution"; that's the controlled production of altered individuals. Evolution is the alteration of the genetics of entire populations, and, if it is to last, to really count on the ledger book of life, it is the alteration of the genetics of entire species—something that appears to happen mostly, though not exclusively, as new species are being derived from old. Evolution is the *permanent* alteration of the genetics of entire species. The most prodigious gene-jockeying task force imaginable could barely scratch the surface of the collective genome of six billion people. Human genetic engineering will continue, instead, in the tradition of eyeglasses, hearing aids, and artificial limbs: addressing the maladies, infirmities, and deficiencies of individuals, one by one. And though such palliatives may well affect the reproductive success of the beneficiary (eyeglasses certainly helped me through to reproductive age), they themselves stand no chance of spreading throughout the vast human genome—and thus being legitimately considered true products of engineered evolution.

Like every other form of life on Earth, we humans have evolved. We have evolved through speciation and natural selection. We have been living in local populations, as parts of local ecosystems, where natural selection goes on all the time. Our species—and those ancestral species before us—basically remained stable, in stasis, for prodigious periods of time. We have always played by the ecological and evolutionary rules.

But no longer: culture now informs the human approach to life, far outweighing, though not entirely eliminating, the older forms of purely biological evolutionary adaptation. Culture has taken us outside the confines of local ecosystems, changing the rules to some extent on how natural selection works on us. These transformations of human existence fit right in with another huge change in human existence: the unhooking, or "decoupling," of sex from reproduction—the essential ingredient for understanding why people have sex.

# Part Three

# The Human Triangle

# EIGHT

# Sex Decoupled

People are obsessed with sex, if not all of the people all of the time, at least most of the people most of the time. Sex sells, and websites with "sex" in the title reputedly get twice as many "hits" as those that don't. More, if it says "oral sex."[1]

Now, it is possible, I suppose, to maintain that all forms of sexual behavior—from straight-out (though "protected") heterosexual coupling, through oral and anal sex, self-gratification, fellatio, homosexuality, cross-dressing, pedophilia, nymphomania, voyeurism and exhibitionism, sadomasochism, and rape—are all spin-offs, just side effects ("epiphenomena") of the urge to make babies. Though this is the dominant theme of evolutionary psychology, I don't subscribe to that, nor, in my opinion, does the average person conducting, to varying degrees, an economic life, a reproductive life, and a sex life.

People are the worst exemplars of the principles of sociobiology. If basic ultra-Darwinism is flawed in assuming that natural selection is simply a matter of genes fighting for position in the next generation, and that such genetic competition drives all of evolution; if vertebrate social systems, including those of our closest living relatives, don't really fit the sociobiological paradigm of "inclusive fitness" and related constructs all that well, we have to ask: How well does evolutionary psychology fare in applying such principles of hard-core evolutionary genetics to humans? As far as I can tell, not very well at all—all the hype and hoopla of finally producing a true "science" of human behavior to the contrary notwithstanding.

In humans, sex is clearly decoupled from purely reproductive matters—meaning that humans don't have sex just to make babies. If social systems in general are amalgams of reproductive and economic adaptations, human society goes them all one better. For us, sex is the bridge that connects the economic and reproductive worlds. There is sex for sex's sake, of course, though a closer look at even the most obvious of examples (such as masturbation) often reveals symbolic meaning that transcends the pleasure inherent in "self-gratification." But most often, sex is clearly tied to the realms of *either* economics or reproduction, and sometimes both simultaneously. This is the human triangle; the simple diagram (figure 6) symbolizes a complex welter of human interactions and behaviors unlike anything else seen in the natural world.

Reproduction is the only one of these three activities that needs the other two to happen—at least successfully. Much about human economic life has little or nothing to do with either sex or baby making. Much about people's sex lives, as well, goes on without much contact with one's economic life—or, of course, with making babies. But reproduction obviously requires sex, and for reproduction to be successful, for the baby

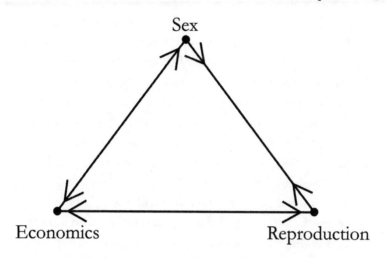

FIGURE 6. The human triangle. Sex is decoupled from reproduction in human life, leading to complex interactions between sex and economics; sex and reproduction; and economics and reproduction—as symbolized by the arrows on the diagram.

to live as a viable economic organism in its own right, requires nourishment, protection, and socialization, which themselves absolutely hinge on the economic inputs of one or both parents—or of surrogates who assume the job partly or wholly in their stead.

## JEALOUSY

Almost all of the rhetoric of evolutionary psychology[2] is devoted to the reproductive implications of one or another aspect of human behavior. For example, patterns of human jealousy, said to be different in males and females, are alleged by some evolutionary psychologists to reflect underlying, genetically based differences in reactions to sexual infidelity thought to have been evolved because males and females face different costs and have different agendas when it comes to making babies. According to this line of thought, human males have biological reasons to seek multiple matings—the better to spread their genes around and thus leave relatively more of them to the next generation. Females, in contrast, are stuck with a nine-month interlude of gestation of (usually) one baby at a time, then months or even years of primary responsibility for care of the baby. Their "costs" of making babies are therefore higher, in terms of time, expenditure of energy, discomfort— and even the possibility of death. Males make sperm at next to no cost and spread them with comparative effortless abandon. Their costs are less (though men do grumble about the costs in time, effort, and money of dating just to get to the point where they can "score").

So jealousy fits into this basic—and at face value fairly plausible—characterization of the different biologically based reproductive agendas of men and women. The argument essentially says that women in the old days didn't care much if the old man cavorted with another woman for pleasure while he was out and about—so long as he brought home the bacon, or whatever meat was available on the Pleistocene African savanna. The danger, according to this scenario, in a woman's eyes arose if the guy became emotionally entangled with that other woman—and started bringing home the bacon to *her*. Guys are said to have been worried about what was going on at home, specifically if any extracurricular mating by their woman while the guy was out hunting might end up with his feeding someone else's kid— the ultimate horror in the tightly wrapped world of ultra-Darwinian thinking. These days, men are thus said to be much more upset by marital infidelity than are their wives, who are supposedly much more worried about emotional infidelity.

I'm not saying this scenario is wrong. It happens, though, that it has been heavily criticized recently by some psychologists who doubt the original data. And indeed it seems that the proportion of men who find sexual infidelity more distressing than emotional straying in their "mates" varies widely from culture to culture: only 25 percent of Chinese men were found to be more distressed by sexual than emotional infidelity, compared with higher percentages of men in other cultures.[3]

This example of jealousy—quite apart from who is right or wrong on it—raises some interesting questions. For one thing, evolutionary psychologists seem themselves to have a hard time deciding whether a particular, supposedly evolved, hence genetically ensconced, trait is still contributing to the fitness of individuals—that is, still functioning as the adaptation it originally evolved for—or whether it was evolved for such reasons in the past and is merely a hangover or even (in the case of rape in some theoreticians eyes) actually maladaptive. The stronger claim would be that evolved traits are still functioning—and therefore still contributing to the fitness of individual humans.

The more relaxed assumption that psychological traits evolved in hunter-gatherer, Ice Age humans, long before the advent of the agricultural revolution and settled existence, but are no longer necessarily affecting the fitness of individuals, is a far slipperier construct. In worst cases, it is the great Pleistocene cop-out—an arm-waving exercise that can't be examined in any rigorous way. Yet even here, there are means of getting at what life might have been like on the African savannas 100,000 years ago and more—as there are still vestiges of hunter-gatherer societies (now almost entirely gone, and none in a "pristine" state, unaffected by contact with postindustrial civilization), thankfully recorded by ethnographers and others, some more meticulously than others. Left, of course, are arguments over what these data might mean.

Suffice it to say, there are ways of grappling with fitness issues arising from specifiable bits and pieces of human behavior, like jealousy. Another problem in applying a genetic-oriented paradigm to explaining the origin and persistence of a behavioral trait is establishing that it is indeed genetically inherited in the first place. Evolutionary psychologists have, as one of their main goals, the establishment of "universal" traits in *Homo sapiens*—the underlying assumption apparently being that anything found to be true of all humans (save the unfortunates who are impaired) means ipso facto that it must have a genetic basis and therefore must have evolved. Anthropologists are equally fond of coming up with exceptions to universal traits, evidently thinking that a few exceptions can invalidate a general rule.

Neither line of argument is decisive or persuasive. But it does remain problematic that the genetic basis of, say, jealousy has not yet been firmly established. This is not to say that you have to know anything whatsoever about genes to say something meaningful about the evolutionary process. Darwin, for example, formulated natural selection without a clue how it is that organisms tend to resemble their parents. It sufficed simply to know that it is so. My long-dead fossils retain not a whit of DNA, yet bear traits that carry over—sometimes unchanged, sometimes further modified—for millions of years, clearly reflecting genetic heritability. It is possible to characterize those patterns as well, and to say something about evolutionary processes—without knowing the molecular details.

Yet critics of evolutionary psychology have a point: there is a striking lack of concern in establishing the genetic basis of the supposedly genetically based traits they say have evolved.

Relative fitness reflects variation within a population of heritable traits. What is the heritability of jealousy? Never mind, for the moment, the alleged male/female differences in jealous reactions to sexual versus emotional infidelity: Does a daughter tend to inherit from her mother her mother's level of jealousy? What genetic effect does a mother's level of jealousy have on the jealousy expressed by her sons? (A reasonable question, unless it should turn out that the allegedly different male and female jealous relations are sex linked—if, for example, male patterns were associated only with the Y chromosome.)

Given the failure to establish the existence of an actual gene for jealousy, a gene for religious susceptibility, and so on (all such kinds of genes have been mooted), it would be important merely to establish that there is genetic heritability of these traits—in the same way that geneticists can calculate the heritability of milk yield in cattle, or coat color and richness in mink.

But heritability is a minefield, of course. Twin studies have done a lot to clarify the influence of genetic heredity and the "environment" in which a child is reared. No doubt about it, personality traits are heritable, to varying degrees. But say we were able to show that there is a positive correlation between a father's degree of jealous rage when he finds his wife has been having an affair and the degree of rage expressed by his son when he, too, finds he's been cuckolded a few decades down the track. Did he inherit the father's level of rage—through a gene with the same "penetration"?[4] Or did he see his furious father rampaging around the house, and was merely repeating the performance?

## Telling Nature (Genes) from Nurture (Learned Behavior)

Ah, nature/nurture again. How to tease them apart? If it is true that human behavior is both learned and inherited genetically, so that the problem for any one instance of behavior is to try to figure out how much is genetic and how much learned—how *boring* we seem to find it. Evolutionary psychologists, paying (often insincere) lip service to learned behavior, would like to be rigorous and scientific and thus want human behavior to be not only largely genetically inherited but inherited with variance—so that selection can mold behavior as adaptations. Nothing, on the surface, wrong with the wish, though at the very least there's a lot of work to be done establishing the heritability of individual bits of behavior.

And I suppose it might be true that there are some "humanists" who still think we are independent from the natural world—or at least that our finest sensibilities are cultural, rather than biological, endowments. But as was pointed out long ago when sociobiology first came along, humans throughout recorded history have known that we are animals,[5] and since Darwin, at any rate, we have known that we have evolved. The possibility that both the brain and our behaviors have evolved along with our bodies is not quite the unexpected, revolutionary shock that first sociobiologists and now evolutionary psychologists often proclaim it to be.

But there is a way, I think, to tease apart the claims of adaptation-oriented, gene-centered evolutionary psychologists and the counterclaims arising from various branches of the social sciences. Tease apart—and arrive at some sense of what is more likely to be the correct explanation. I am not talking here about the degree to which offspring resemble their parents, and the degree to which genes versus learning determines whatever level of heritability might be demonstrated. Rather, I am talking about the more general kinds of assertions that evolutionary psychologists tend to make—for example, that men and women everywhere have different levels of jealous reactions to sexual versus emotional infidelity: anything about human behavior that leads an evolutionary psychologist to postulate that the behavior arose to maximize the reproductive success of individuals, male or female.

The question boils down to this: Is the competition for reproductive success the main reason people do what they do? Or is human life an interplay of sex, economics, and reproduction—with reproductive success, whether in the Pleistocene or now, playing a role, but perhaps more like a bit part than the major determinant of what we do? How can we know?

I'll actually borrow a page from ultra-Darwinian biology to devise a litmus test to help us evaluate whether competition for reproductive success or some other, cultural process—learned behavior—is more likely at work as the ultimate raison d'être of any particular aspect of human behavior. At issue in the 1960s was the challenge posed by the concept of "group selection"— where a form of evolutionary selection was considered an additional possible mechanism of evolutionary change—additional, that is, to good old, straightforward, natural selection acting among individuals within a population.[6] *Natural* selection works among individuals of the same species within a local population. *Group* selection, in contrast, deals with the relative success of groups (demes) within species.

It was the biologist George Williams, in his famous 1966 *Adaptation and Natural Selection*, who first showed the way to distinguish between natural and group selection. He said, in effect, that if both forms of selection were operating, and were both producing evolutionary change in the same general direction, there would be no way to tell them apart and to determine that group selection was acting alongside of natural selection.[7] What we need, Williams said, are examples where the two forms of selection are working at cross-purposes. Surveying the literature, he could find only one example that satisfied him: Richard Lewontin's example of the *t*-locus in mice.[8] Thus Williams concluded that the evidence for group selection in nature (or experimentally) was exceedingly rare, leading to a hardening of views of individual genetic selfishness in selection.

In the same spirit, we might look for evidence of processes that run counter to—even override—the "normal" expectations of selfish-gene, fitness-based explanations of the origin and maintenance of human behavior. Such examples would be good evidence for other, expectedly cultural forces at work shaping human behavior. An example, so clear that it borders on the trivial, is China's one-baby-per-couple official policy, supposedly enforced only on the Han people (the numerically dominant population in China). This is a political rule, based on overt concerns with China's burgeoning population and its relation to economic productivity, that obviously plays hob with the normal, biological "fitness" of both individuals of a married couple—not to mention older customs (see chapter 10) that themselves were biasing the reproductive output of individuals belonging to different sexes, classes, and even regions (e.g., rural versus cities).

## HIERARCHIES IN HUMAN LIFE

Anthropologists have long argued that the social "infrastructure" dictates in large measure what people do in their societies.

They have also made the stronger claim that any aspect of human behavior documented in a society—especially, though not exclusively, in the area of what is considered normal and acceptable behavior in sex and baby making—is not written forevermore in the genetic architecture of individuals. They point to the myriad cases documented when individuals are readily socialized into other beliefs and norms of behavior.[9]

The Chinese population control example raises the additional point that the cultural "infrastructure" frequently takes the form of hierarchical control—dictates from above in the power structure. As we saw in the first section of this book, organisms, by sheer dint of their reproductive proclivities and their economic behavior, find themselves parts of larger-scale systems, which are themselves hierarchically arranged. The same can be said—in spades—for human social systems, especially human *economic* systems.

In postindustrial society (itself based on the rise of city-states after agriculture made possible, even demanded, a settled existence), the proliferation of purely economic hierarchical systems is mind-boggling. Americans live in neighborhoods, which are parts of towns or cities, themselves parts of counties, then states, then finally the whole United States. And, of course, the United States is but 1 of 3 North American countries, and these are but 3 of the 173 major countries in the world. The political hierarchy is an exact analogue of the economic hierarchy built and occupied by the rest of life—the hierarchy of populations, local ecosystems, and larger such units. Because we no longer live in a local ecosystem, we have developed our own regional economic system, the geopolitical hierarchy of nation-states, in its stead.[10]

But increasingly in the modern world, we do not work where we live.[11] Here, too, we encounter a maze of hierarchies. A person might work as an accountant in a department within a division of a corporation, which itself is a part of a sector of trade (e.g., one oil company out of many), but it is also a part of the regional economic structure (one tax-paying, employing business out of many in, say, New York).

Back at home, a worker in the city may also be a member of a local church, which tends to be affiliated with other churches of the same denomination in the region (perhaps overseen by a hierarchy of prelates), and often has national or even international affiliation. Much the same is true of membership in clubs like the Kiwanis, Elks, or Rotarians. And the same holds for kids going to school, or for their participation in scouts. These are all economic activities (though often not moneymaking), but nonetheless outgrowths of the material, even spiritual, yet decidedly *not* reproductive, side to life. To think of the economic side of human life is automatically to think in terms of hierarchical systems—many, many hierarchies, all at once, as opposed to the single, local economic system that a bird or a plant belongs to.

Though many of these hierarchies are like the economic and reproductive hierarchies of all the rest of biological life (i.e., consisting of nested sets of parts within wholes, such as organisms within demes within species), they are also, in many instances, *control* hierarchies, such as the organizational/power control of the U.S. Army, where a control flows from the few at the top down, through a series of ranks to the many buck privates. Control takes many forms: laws or rules (as in the China population control policies) are the easiest to spot. More subtle are the "social norms," the unwritten expected rules of behavior—amply documented by anthropologists for a variety of non-Western cultures, but harder to see by individuals living in their own cultural milieu. We know what we "ought" to do, in terms of an almost endless series of behaviors. In suburban New Jersey, at any rate, we are expected to keep the grass cut and the house in good repair (at least as seen from the street or by the neighbors).[12] We know, too, what the unwritten rules are regarding extramarital sex—rules that nonetheless at times seem made to be broken. And though Americans routinely deny that there are any rules (written or unwritten) governing our reproductive behavior, the availability, or not, of contraceptives and abortion—still hot political topics—suggests otherwise.

Three examples out of an endless array of possibilities, but three that show that social forces—generally from above—dictate in large measure what we do, or at least how we are supposed to behave. These examples include (1) economic behavior (lawn mowing), (2) extramarital sex (sexual behavior), and (3) reproduction itself. Never mind for the moment that there are *possible* connections between extramarital sex and reproduction (see chapter 9), economic behavior and sex,[13] and economics and reproduction. The point is that these three reasonably separable categories of human behavior are all subject to social regulation—from above, from norms evinced by peers, and from (perhaps most tantalizingly of all) the exigencies of belonging to one socioeconomic class or another. The category of socioeconomic class offers a particularly rich set of contradictions to simple expectations based on the assumption that all of us are out to maximize our reproductive success, and that the chase to spread our genes underlies and essentially dictates the fabric of society—in Pleistocene hunter-gatherer bands, as well as on the streets of New York.

# NINE

# Up Close and Personal

## *Sex, Power, Money, and Babies*

One thing that makes the television soap opera *The Sopranos* so appealing is the stark contrast between the violently sociopathic life of a mobster and his perfectly conventional upper-middle-class home life in the suburbs of northern New Jersey. There they are: Tony, his wife, Carmela, their daughter, Meadow, and younger son, Anthony Jr. ("AJ"), living in a large, fairly new McMansion, replete with swimming pool and vanity SUV. There's even a shrink in Tony's life as he wrestles with the conflicts of living in two incommensurate worlds.

The four members of the Soprano family are stereotypically American: mom and dad, sister and brother—as in the old cereal box scene of a surreally calm mother pouring out cornflakes for her brood while mild-mannered dad looks on with a goofy grin on his face. There it is: food, love, procreation—with

the kids about to go off to school and dad to his work as accountant, paleontologist, or hit man. The families—especially the
"nuclear families" of parents and kids—are the smallest of
human social systems. They are also the only units that combine all three: economics, sex, and baby making.

Most families these days do not conform to the stereotype of
dad, mom, and a couple of kids—and probably never did. Fifty
percent of American marriages end up in divorce, and single-
parent households are rife. Then, too, many marriages remain
childless throughout their duration, whether by choice or infer-
tility. The assumption that all humans are driven to make babies
to leave copies of their immortal genes to the next generation
is nicely countered by the existence of same-sex liaisons and
heterosexual couples—such as the famous Madison Avenue–
named DINKS, "double income no kids" unions in which husband and wife (or both unmarried partners) pursue careers in
the world—and a home life free of the responsibilities and costs
of raising kids. And though it is true that some homosexual couples may indeed raise children—sometimes the offspring of a
previous marriage of one or perhaps even both, sometimes
through artificial insemination, sometimes collateral kin, and
sometimes pure adoption of totally unrelated children—for the
most part, gay and lesbian couples do not have children,
whether their own or someone else's, in their day-to-day lives.
Leaving aside the nature/nurture issues of the basis of homosexuality (i.e., whether homosexuality is ensconced in the genes or
is instead a behavioral choice, or a mixture of both), gay and
lesbian couples can procreate and rear children should they
want to, and most appear not to.

So not all families are alike by a long shot. Even if we restrict
the term "family" to the kind where kids are a part of the scene
(or were—though it increasingly has seemed that they never

will leave, most eventually do), families differ markedly in size, age spacing between children, ratios of girls to boys (not always a matter of chance), and patterns of allocations of resources to the kids.

All that being said, it is obvious that the family in the restricted sense of mom, or dad, or mom and dad—and the kids—reflects a pretty tight integration of economic, sexual, and reproductive behavior. The Sopranos are a good case in point: Tony is the sole breadwinner—not perhaps the most common situation in married family life in northern New Jersey these days, where the present generation typically needs to have two incomes to maintain standards equivalent to the suburban life today's parents knew as children themselves. Carmela's economic activities are, however, prodigious, for she is the one who maintains the house, shops for and prepares the food (except for the cannolis that Tony sometimes brings home), tends to the kids, pays the bills—and badgers Tony for an economic plan to cover herself and the kids "should, God forbid, anything ever happen" to Tony.

Then there is their sex life, depicted as off and on, and easily disrupted by depression (Tony's), household spats (often, but not always, over the kids), and other downers. Tony conducts a sex life outside the house, whether vicariously, surrounded by bare-breasted women at the Bada Bing, or directly with his current girlfriend. Like everything else in Tony's life, the fun outside-the-house sex life also presents its problems. Carmela, meanwhile, literally flirts with extramarital affairs, but so far, at least, has not acted on them. Meadow starts having sex—intercourse—in college and takes measures to protect herself. Her younger brother, AJ, has yet to show much of an interest in sex, though (as I write) that seems about to happen.

And, of course, Tony and Carmela had sex at least twice to

produce their children. The money goes to keeping hearth and home, body and children, alive and to sending them to college. The kids, more often than not, are the apples of their parents' eyes, and seem somehow to be worth all the aggravation and financial outlay.

There is thus a close association between sex, baby making, and economics, nowhere closer than in a nuclear family. The Sopranos might be fictitious, but they reflect reality very well indeed.

## ALL IN THE FAMILY

Long before evolutionary psychologists, in their search for human universals, arrived on the scene, other people had been wondering how universal nuclear families really are. The anthropologist Melford E. Spiro once asked just that question; he thought that Israeli kibbutzim—where children are raised communally by surrogates, with most of the jobs of child rearing usually handled by biological parents now taken over as a task by community specialists—to be one of the comparatively rare examples of non-nuclear-family structure, the exception that "proves" the rule. Other anthropologists, of course, disagreed with the claim that the kibbutz represents a significant departure from the nuclear family.

More recently, we have reports of the Na people, one of China's ethnic minorities living in Yunnan Province.[1] The Na know no marriage; instead, "visits" by males to the bedrooms of females go on supposedly ad libitum. Children know their mothers, but do not know the identities of their fathers. Thus the typical household arrangement seems to consist of women and their children—including daughters old enough to have "visits"—and male relatives of the women in the house. Only on rare occasions, reportedly, will a man used to having visits with a woman in a house actually move in—only when there are not enough men already around the house to perform the essential economic labors that keep things going.

So "marriage" and the "nuclear family" are not completely universal, but they are nearly enough that exceptions are few. And even when some apparent exceptions crop up that are sufficiently credible and compelling that some analysts, at least, are willing to say "this is not the usual way of doing things," the basic integration of sex, reproduction, and economics is still present in whatever social structure is there in its stead. There is a mix of sex, reproduction, and economic life in kibbutzim— bonded-pair sex, extramarital sex, baby making among the married—but the nurturing is now expressly recognized for what it is: an economic activity directed to the physical (and, presumably, emotional) well-being of the children, conducted by specialists as part of their kibbutz work, rather than by the biological parents.[2]

The Na also present a mix of sex, reproduction, and economics in their lives—though with the focus in the reports so heavily on the sexual "visits," less seems to be known about the economics of village life. Yet males do in fact seem to be involved in the daily economic pursuits of household compounds, there is plenty of sex, and babies are reared by their mothers in an economic environment in which both men and women of the household participate. In a kibbutz, you get married and have kids, but someone else raises your kids. With the Na, you don't get married, but the children are raised in a home that is itself an economic cooperative among close relatives— and the occasional male who is allowed to move in.

However much such alternatives deviate from the nuclear family, they all present the familiar theme of interlocking economics, sex, and reproduction. How did this tight integration of economics, sex, and baby making come to be?

## THE SEX-FOR-FOOD, FOOD-FOR-SEX SCENARIO

How did sex come to be decoupled from reproduction, and at the same time to be so closely tied to economic human behavior? Recall that bonobos, the most humanlike of our closest living ape relatives, indulge in a great array of sexual activities, much of which have nothing to with pregnancy.[3] Couplings (often missionary style) take place at all ages, female–female genital contact is common, and a variety of other noncoupling sexual behavior goes on between all ages and gender combinations. The decoupling of sex from reproduction obviously started a long time ago and, like other aspects of biologically imbued human behavior, goes way back in primate evolution and is not evolved only within our species or even our closest evolutionary ancestors back on the Pleistocene savannas of Africa, as evolutionary psychologists commonly suppose. Frans De Waal and other close observers of bonobo behavior believe that much of the flamboyant and incessant sexual behavior of bonobos has far more to do with reducing tensions, such as those arising over competition for food. Sex among bonobos is not only decoupled from reproduction; it seems also to have overt economic implications.

Human females are continually sexually receptive and can conceive twelve months of the year, albeit only for a few days each month. Though there may well have been some reproductive adaptive value for being able to make babies year round, I have no idea what that advantage would be. All but the largest families on record could just as easily have been produced with a single seasonal period of estrus, given the nine months it takes for a human fetus to develop fully.[4]

It makes more sense to think that access to continual sex in humans evolved not for reproductive purposes but rather as a kind of "food for sex, sex for food" arrangement[5] that led to the creation of more stable pair bonds than are found, for example, among bonobos. It's an arrangement with clear reproductive advantages, since such pair bonds would presumably have led to a more protracted and stable "family" life, in which economic necessities were looked after by both parents and care of offspring particularly (if not exclusively) was the domain of the mother.

Humans are notoriously helpless at birth; walking doesn't come for about a year, and speech usually takes two or more years. Kids simply cannot fend for themselves until, as a rule, well into their teens or (in some situations in some societies) even later. To provide the economic support for such a labor-intensive job as raising a human child, stability of the support system is crucial. Sexual gratification is the carrot on the stick (so to speak), not just to keep the guys around (as has most commonly been supposed, since guys are supposedly the main bread-winners) but also to keep women interested in the arrangement. Obviously, women often work while in the midst of their child-bearing careers. In hunter-gatherer societies, women collect much of the food (chiefly plants) and tend to most of the other domestic economic chores, while the men hunt for animal protein. Hunter-gatherers, of course, are not infallible guides to ancestral hominid life of the African savannas, but theirs are the only preagricultural societies still with us (though just barely), thus our only living signposts to what human behavior might have been like prior to the agricultural revolution that started ten thousand years ago.

So food for sex has deep implications for successful baby

making. But food for sex also means sex for pleasure rather than just sex for babies. If food is an economic commodity, so now is sex, as any item in a barter arrangement is by definition something of value, something that can be bought and sold. For the first time in the 3.5 billion years of life on Earth, sex has acquired a use and a meaning that transcends the pure production of babies.

It is thus not the reproductive imperative, but rather the diffusing influence of taking sex away from its strictly reproductive function, that creates the human triangle and really drives human existence. Sex is now free to have its own life in human affairs—a life that exists in and of itself and has special "relations" with the economic side of human life. And, oh yes, we still have sex to make babies.

## Sex for Its Own Sake

My personal favorite theory for why there is sex at all—and why although the majority of us prefer partners of the opposite sex, some of us prefer same-sex partners—comes from an ancient Greek story recounted in the playwright Aristophanes' comedy *The Clouds*. We were all doubles once (so goes the story), physically connected to our twin half, who was usually, though not invariably, of the opposite sex. Some epochal event severed all the doubles of the world, and now all of us spend all our time running around looking for our double—which means most of us are looking for someone of the opposite sex. But sometimes, some of us, because our double was of the same sex, are looking for *that* person.[6]

Not right away, however. Sexual maturity comes after a period of infancy, and whereas sexual latency is there from early infancy, overt sexual behavior is generally muted until adolescence. Adolescence comes at different times for different sexes, is changeable in history, and varies between cultures. But just ask a young American teenager whether life is different from what it used to be before his or her hormones started to rage. It is very much as if old August Weismann's distinction between the germ line and soma were played out in striking, very personal ways as a kid grows up.

Then sex intervenes. For almost everyone, this is a shock. Priorities change. On the plus side, there's the thrill and pleasure of surging hormones, of discovering orgasms—usually, though not invariably, on one's own, abetted only by gossip from friends, and soon pulsating to the ubiquitous sex stimuli that were there surrounding them as preteen-agers all along. There is the heady success of being attracted to someone else, maybe even of becoming "popular" and doing pretty well from the get-go in the embryonic phases of sexuality. And, last but not least, there's puppy love.

But just as often there is utter doubt and uncertainty. Rejection both real and imagined—with pure fear of rejection itself, or simply the feelings engendered by unfamiliar hormones. Worries about acne, weight, breast size—a million extra worries that simply hadn't clouded life until sex came along.

If ever there were a graphic example—the literal proof—of the fundamental apartness of one's simple existence, one's fundamentally *economic* life, from one's sexual and (possibly also) reproductive life, this is it: human adolescence. There you are, a young human being, eating, sleeping, defecating, urinating, breathing, growing. Sex only vaguely intrudes from time to time on your consciousness—and though you might be acting as a surrogate to help raise younger sibs, baby making per se is something, like being a fireman sometime, that lies some vague and distant time in the future. We can see and feel in our own selves the practical consequences of that germ-line/soma distinction that Weismann drew so long ago: all organisms live both economic and reproductive lives. And nearly all live at least for a while as purely economic entities before they "mature" and start making babies on their own.

In the United States, girls tend to begin menstruation and to develop breasts and pubic and underarm hair on the average around the age of twelve, and boys somewhat later (again, on the average) start growing facial, pubic, and other secondary sexual hair and having far more erections than ever before—erections that now can lead to explosive ejaculations. Eighteen- and nineteen-year-old boys are said to have an erection on the average every ten minutes, think explicitly about sex at least that often, and produce sperm in quantities their fathers can barely remember. In contrast, there is no comparable physiological or psychological sexual peak for girls, since interest in sex continues, even strengthens, in women up through the start of menopause.

From a purely physiological point of view these are, or should be, the prime baby-making years. Frequently, of course, they are not, as schooling and early career development for both young women and men—in the middle, upper middle, and upper classes—more often than not intrudes. It takes an economic life to support a family—a reality that has much to do with relative success in spreading genes in all human societies, and a theme that also is closely tied to the human population explosion that is threatening environments, ecosystems, species, and human life, virtually everywhere on this planet.

So why is adolescence typically as painful as it is thrilling? Barnacles settle out as larvae, metamorphose on a bit of shell or piece of rock, and feed and grow until the time comes when their gonads ripen and they start making eggs and sperm (many barnacle species are hermaphroditic). There is a metabolic cost to sperm and egg production; some of the energy derived from food intake now must go to this new activity. But most barnacles seem to be able to absorb this cost and go on living, not overly traumatized by the sudden onset of sexual maturity and the additional burdens of baby making.

Why, then, do humans find adolescence so typically rough? And not just American adolescents but kids virtually everywhere? Why are there elaborate rituals, like male and female circumcision, or shutting girls up when they first menstruate (or even throughout their entire lives as menstruating women)? Why, if it is no big deal for barnacles (and perhaps not even for bonobos, which show a rudimentary form of disentanglement of sex and reproduction) to simply take on the added functions of sperm and egg production, mating, and perhaps some forms of child nurturing, is it such a big deal for humans?

The answer almost surely is that sex *means* more in the average human life than in the lives of any other sorts of organisms that have ever graced the earth. Sex is fraught with meaning, from the hidden symbolic to the explicitly overt. As the only truly conscious animals on the planet, we are naturally in a position to invest meaning in anything that strikes us about our lives—and that is surely what we have done with sex.

Sex has come to symbolize personal worth and personal power. It ranks right up there with money in that respect—and sometimes even higher. As we shall soon see, the two are sometimes inseparable.

## SEX FOR PLEASURE, AND *NOT* FOR MAKING BABIES

Adolescent sex is not generally about baby making. Yes, there's a lot of out-of-wedlock pregnancy among teenagers in the United States, and in many other places around the world. In some circumstances, teenagers deliberately do seek pregnancy and babies to enhance their social and economic status. For the most part, though, babies are as unwanted as the social diseases that are still rampant. Condoms are routinely sold in men's bathrooms in bars and restaurants around the United States—marked "sold for the prevention of disease only," a sop to the ambient cultural forces (read conservative Protestant and Catholic interests) still adamantly opposed to birth control. In truth, condoms are sold in such venues every bit as much for the prevention of pregnancy as for the prevention of the spread of syphilis, gonorrhea, and now AIDS.

The connections between sex, power, and baby making run deep, and it is especially difficult to tease apart the motives and forces acting on any person's behavior—be that person an adolescent, young adult, mature adult, or older member of society (terms themselves defined only partly biologically and, in any case, differently from culture to culture). Teenagers are notorious for following the perceived dictates of their immediate peers first and foremost, and those dictates are further shaped by standards of behavior in other groups of teenagers, increasingly easy to find out about through electronic media. (Think of Tipper Gore's crusade against what she and many others perceive as encouragement of bad behavior encapsulated in the lyrics of rock, hip-hop, and rap recordings.) Coming in last in the competition for effective influence on the behavior—sexual and otherwise—of adolescents are the norms preached by their parents, schools, and churches, should any of these categories of supervening power be present in their lives.

So it is difficult to look at the "typical" sexual behavior of anybody, let alone of any group of people defined by age and sex, and explain exactly why they do the things that they do.

Biology—rampant hormones—is obviously a strong motivating factor in teenage sex. But so is peer pressure (What? Still a virgin??), and I am aware of only a very few reported instances where making babies per se is part of the social mix that underlies sexual exploits (for example, in the boasting of out-of-wedlock fatherhood of some inner-city young men in the United States).

Tough as it is to separate larger-scale cultural forces, especially the all-important defining factors of economic (class) status, from the biological and psychological factors underlying everyone's sexual behavior, from adolescence into old age, there is still plenty to be said about individual sexual practices that are not overtly connected to either economics or the production of offspring.

Masturbation and bestiality are obviously forms of sex that cannot result in pregnancy. So are the various forms of "sodomy." These sexual practices are pursued strictly for pleasure—and sometimes, of course, for money. The list of sexual behaviors that have absolutely nothing to do with baby making is very large and potentially endless, limited only by the bounds of the human imagination and anatomical realities—which is to say hardly limited at all.

## SEX AND POWER—AND NOT BABIES

Sex has a lot to do with one's self-esteem. There is a common desire to associate with someone glamorous: for men, a beautiful woman; for women, perhaps a handsome man, but perhaps as well an older, powerful man—someone with lots of money, prestige, *power*. Even people who are not especially motivated to seek out a chain of progressively more glamorous or powerful partners are nonetheless not uniformly immune from consuming the often tawdry details of famous people who publicly do live such lives.

For men, it may well be that simply proving, over and over, that you are still able to attract a desirable woman leads to so many affairs and divorces. Even if it were strictly true that the initial urge to find and "acquire" an attractive woman was for the express purpose of setting up a nuclear family arrangement and having children, one more fling at handing down those precious genes, the repeated search for the same heady feelings of infatuation, and the often desperate attempt to prove that you can still make it in the competitive world of beautiful women— all this is for the most part *not* about making still more babies.

Affairs, for either sex, are rarely about gene spreading and baby making.[7] Extramarital sex, whatever its complex motivation (trouble at home, perhaps trying to create trouble at home, or just looking for excitement), seldom leads to the production of still more children.

Serial marriage is another matter. Often, for both men and women, periodic divorce and remarriage is about acquiring "trophy" spouses. A woman dumps a guy to get another one richer or more famous. She may or may not give him children, either as part of the "deal" or perhaps because she wants to, regardless of his intentions. On the other hand, if a guy dumps his wife to marry a woman younger than the previous wife, that can be purely all about trophy collecting, but kids do often enter into the picture.

High school is the place where all the accoutrements of power through sex are first explored, in the form of connections with the economic world. The cute little Honda, or the bright chrome muscle car (depending on time and place, of course), whether or not bought by mom and dad, are status symbols that fall neatly into place. Not that preadolescent kids can't be snobs to one another about their possessions, but women and fast cars tend to go with one another in a guy's mind like ham and eggs in a somewhat more prosaic context. "Cock cars" need not be limited to old Corvettes, nor are they limited in their appeal to the very young. It was in the pages of *Playboy* that a full-page cartoon depicted the stages of a man's life, from youth to old age, through his cars: a banged-up runabout for starters, a new, modestly small car at marriage, a station wagon when the kids come along, and progressively bigger cars as he moves up the corporate ladder—until, at the pinnacle of his career, he becomes chairman of the board and buys an expensive sports car the size of the car he started out with. Big smile in his face, a cigar clenched in his teeth as he races down the road, the air blowing what hair he has left.

Power. What person you are with, what car you drive (could be construed as purely economic power, but there is that connection with sex), how long your penis is. Women worry about how slim they are and how curvy. Women do tend to worry about the size of their breasts, but apparently not nearly to the degree to which men tend to worry about penis size. The Internet now has as many ads for penis-enlarging products as for lower-rate mortgages and computer printer ink cartridges.

And now we have Viagra, latest in an unbroken string of "male potency" nostrums that extends back to the dawn of human history. Viagra has been an amazing hit. The former senator Bob Dole, one of America's current crop of elder political statesmen, has enthusiastically and very publicly endorsed this product.

Viagra is emphatically not about making babies. It is selling like mad (and through the Internet) mostly to men in their fifties, sixties, seventies, and eighties—men whose most vibrant days are long past, but men with memories and with wives or girlfriends who still want sex lives. Sure, some of them might have younger wives where the issue is perhaps more pressing, and some of these younger wives may be intent on having a baby or two. But most of Viagra's use is of the "à la recherche du temps perdu" variety of experience, a recalling of youthful vitality, strength, potency—power. Surely a significant amount of the pleasure Viagra brings comes from the simple fact of erection as much, perhaps, as from the actual sexual experience that presumably follows.

Women, too, crave sex far beyond their reproductive years. Menopause, the official and internally very noticeable end of the reproductive years, can be a physical challenge as well as a manifestation of aging and decline of power that is often emotionally difficult to deal with. A silver lining not missed by many women, though, is that birth control devices are no longer required (though concerns about disease never go away). But no more pill, and never again the specter of abortion or RU-486 (for women with access to such modern chemicals and the desire to use them). Whether or not one has reproduced, once the fertile days are over, sex can be, and usually is, still very much in the picture. Sex symbolizes power and success to American women as much as to American men, perhaps even more so.

But there is a downside to everything, and the psychological dangers of having so much of one's feelings of success, well-being, and power wrapped up in sex is that there is an awful lot that can (and routinely does) happen in life that upsets the applecart.

For one thing, sex itself is risky. There are the risks of disease. There is the risk of pregnancy: for when pregnancy raises the specter of loss, the costs are immediately seen as too great. The continual stories of babies dumped in bathrooms and garbage disposals—babies born of upper-middle-class as well as of poor mothers—are graphic reminders that even when there is plenty of money for child support, the reality of a child's existence can represent such a drastic threat to the future already planned that the child simply has to go.

Even when babies are wanted—at least in the abstract, "eventually"—their arrival is commonly planned (sensibly enough) around the economic needs of career building and establishment, and often, if conception occurs prior to the planned moment, that pregnancy will be terminated for reasons of convenience and economic necessity. Sometimes, of course, those delayed pregnancies never do happen; the biological clock simply runs out, or career turns out to be more compelling than baby making.

The connections between sex, health, and psychological well-being run deep. At the physical level, it is notorious that women athletes who train rigorously will sometimes stop menstruating; sex can go on, but there is temporarily no chance of becoming pregnant. It is very much as if excessive stress to the economic side of one's body comes at the expense of the reproductive side.

Male athletes do not face that issue, but just as sex can lead to drowsiness in both men and women (even postcoital "tristesse"), there is a persistent sense that sex successfully completed results in a temporary loss of power. While some coaches of male teams have been known to encourage at least some members of their teams to enjoy as much sex as possible,[8] more of them have reportedly tried to temper their players' sex lives so as to save their strength and concentration for the game or event that lies ahead.[9]

But if sex helps convey a sense of power, anything else that brings about a loss of power, whether real or imagined, can have devastating effects on anybody's sex life: the vector, in other words, is reversible. Illness can do it, even if the illness itself does not technically prevent sex from happening. Depression, whether of internal origin or caused by external events—or the more usual combination of the two—is a notorious anti-aphrodisiac.

This is especially true when it comes to the links between one's economic and sex lives: loss of job, career reversal of any sort, any kind of financial disaster can very quickly undermine one's sense of well-being and power. And there, usually, goes the sex life right along with financial security. A sort of cultural analog to the depression of reproductive functions (if not libido per se) in overexercised female athletes, economic loss at the level of individual human beings is at least as graphic a symbol of loss of power as is sexual dysfunction—and can lead to loss of sexual appetite every bit as quickly as economic success can stimulate one's sex life.

## Sex, Money, and Power—but No Kids

The connections between economics and sex run very deep. As we already noted, both are commodities (as babies can be as well). When sex leaves the house and enters the world of economics, it immediately becomes difficult to distinguish the needs, drives, and behaviors of individuals pursuing their sex lives for fun, power, and profit from the cultural dictates arising from social norms and economic exigencies of class. Though it can be fairly easy to separate the acts of individuals from the larger-scale social patterns of infanticide, rape, and slavery, the more subtle and arguably less destructive acts of individuals indulging in sex at the office, visiting prostitutes, or consuming some of the vast quantities of ubiquitous pornography have both individual motivations and social overtones (including social sanctions that at once discourage and abet such behavior) that are frequently hard to tell apart.

Take office sex. No doubt the office is a great place for conducting an active dating or sex life, and much of this behavior involves the core search for a mate suitable for pairing off and settling down and, in the case of heterosexual couplings, eventually marriage and production of offspring. Yet, despite the great inroads women have made in American economic life—to the point where many more women than ever before hold positions of power in corporations (including that of CEO) and in the professions—"sleeping up the ladder" has seemingly not been abandoned as a strategy for success in the economic world. Sex will not stop being a medium of exchange as long as there is a perceived payoff in terms of job advancement. Men, on the other hand, seem to pick on women with lower status at the office (not difficult in any case, since women remain at lower levels of remuneration and status in the workforce), so when already married males pursue women at work, their motives seem to be largely the power and fun of an extramarital fling, possibly enhancing their reputations among peer males, rather than direct economic gain—or the production of children. But all permutations and combinations take place, and women higher up the corporate ladder are also known to select a male of lower status.

Nowhere is extramarital sex more closely linked to power and even glamour than in the realms of show business and politics. The extremely visible attempts to bust Bill Clinton for his escapades with women as governor of Arkansas and, especially, as president of the United States (where his article of impeachment boiled down to his allegedly having lied about his sex life, rather than to the fact he was conducting one) made the United

States a laughing stock of the rest of the world (especially in Europe, where such behavior is recognized as the norm). When François Mitterrand, the president of France, died, both his wife and his mistress (and assorted children) attended the funeral. As a young man, Bill Clinton was inspired by John F. Kennedy, whose extramarital sex life was discreetly ignored by the Washington press corps, though well known to all. Gary Hart was another ambitious politician who thought it was all right to engage in various sexual high jinks; like Clinton, he mistakenly assumed that the same rules that had applied to powerful politicians in the recent past (and, stories have it, off and on pretty much throughout the history of the American presidency) continued to apply even when the political and "moral" climate shifted toward the right. If it weren't so traumatic, it would be downright hilarious that some of Clinton's biggest political enemies on the Hill, including House Speaker Newt Gingrich, were also found out to be conducting clandestine extramarital affairs. No stranger to hypocrisy, American political life nearly absorbed the impeachment of a sitting president while his enemies were carrying on much the same way: Kenneth Starr's beady gaze caught Clinton with his pants down and produced a parade of women eager to tell of similar previous encounters, while the others thought they were safe. The whole messy episode shows how closely associated sex is with power, especially with the sort of heady power conveyed by high positions in national politics. Clinton's "crime" was in misjudging the political climate, far more than in being a unique departure from the typical shenanigans of high-power political life.

Then there is Hollywood. However straitlaced, goody-two-shoes the American public seem to consider itself, it has a collective fascination for the tawdrier, sexual side of our movie stars. I have no idea whether the starlet sleeping-up-the-credit-line days are over (it seems unlikely that those days will ever be forever gone); in any case, starlets and their male counterparts are endlessly, publicly active. It's like high school gone mad. You can't go through a supermarket checkout line without seeing sexy tabloid headlines (not, of course, unfailingly accurate). If there is a general human triangle of economics, sex, and reproduction, here we find triangles of the old familiar variety— someone is always out screwing around with another, at times more famous partner, leaving the purportedly confused and certainly betrayed spouse at home (or off on his or her own escapade), there to suffer public humiliation.

It's all about publicity, for it has long been a public relations axiom that public notoriety for whatever reason is good for the career—so long as they spell your name right. Sex is probably the easiest way to keep your name in the gossip columns. Much better than drug busts, and less of a threat to your overall health. It can even help overcome flops and bad reviews.

Nor is Hollywood the only locus for nonstop sexual news. You might think that a hit record or two is the ultimate goal of a rock musician, but most bands are still heavily into sex, drugs, and rock and roll—to the point that no tour is going well unless it has its full complement of groupies for postconcert partying. This is not a new phenomenon: when big-band "swing" music was the popular music of the nation (especially from the mid-1930s into the early 1950s), musicians on the road lived a life fraught with sex, and often with drugs, and with the boredom and loneliness that comes to the lives of all who are forced to spend much of their days away from home.

Sports, too. For every story of a (recently retired) Joe DiMaggio hooking up with a Marilyn Monroe (a famous celebrity marriage that ended in divorce and no children), there are countless episodes of athletes, married or not, with active sex lives in every port of call.[10] Though it must be admitted that efforts are here generally made to stay out of the newspapers (not always successful—the famous Copacabana brawl involving Mickey Mantle, Billy Martin, and other Yankees included women as well as booze and other male brawlers). Yet it is the glamour and power that is the lot of famous athletes that gives them access to women (and a host of other things that threaten to mess up their lives). Let us never forget that Wilt Chamberlain—he who boasted that he'd had ten thousand women as sexual partners over his sexually active lifetime—was an incredibly successful athlete, a famous basketball star who once scored 100 points in a single game.

Sex is about pleasure. Sex is about baby making, some small percentage of the time. But sex is also about power, symbolically and even financially. Sex confers and conveys power—and power in the form of fame, fortune, or simply a sense of well-being in the more humble lives that most of us lead. Sex and power feed off one another in an endless dance.

But sex is also for sale.

## Love for Sale: Sex as a Commodity

The world's oldest profession? That was probably the hunting-gathering amalgam that kept people alive as ancestral humans, though if there is anything to the "food for sex" scenario of early hominid pair bonding, sex has been an economic commodity since the earliest days of hominid existence.

But if by "profession" we mean the things people do in exchange for material goods (food, manufactured items, eventually money) and services after the invention of agriculture and the sedentary life that came with it, then there is little reason to doubt that prostitution—sex for sale—was there from the get-go. Amply documented in the Bible (which is, after all, one of the oldest written records shedding light on what life was like thousands of years ago), sex for shekels goes back to a time long before Mary Magdalene's.

Sex for hire is no laughing matter. For every college girl who works at a topless bar for kicks and spending money, and maybe throws a few tricks more or less as a lark, there are hundreds of thousands of women who sell their bodies as the absolute last resort for their survival—and frequently their children's survival as well. Even worse (if possible), there is the rampant sexual slave trade—ranging from families selling their daughters to work in the brothels of Southeast Asia to the importation of European, African, and Asian women to serve as sex slaves right here in the United States. These are sweeping social issues, and especially because they often involve the production and sup- port of children, I'll get back to the dark social-issues side of sex for hire in the next chapter. For the moment, though, it is worth taking a look at the commercial side of sex strictly for the con- nection between money and sex—to the patent exclusion of baby making, at least insofar as the sexual partners are con- cerned.

The sex industry is worth billions of dollars a year to the American economy alone. Prostitution might be the oldest component of this ancient business, but pornography is proba- bly the most lucrative.

Prostitution, it seems clear, will never go away. Beyond the hard fact that there will always be women who need to support themselves, their kids, or, sadly, their drug habits, there is the equally hard fact that there will always be men willing to pay for sex. The anonymity of it—the very real sense that paying money removes any sense of obligation and thwarts any propen- sity toward emotional entanglement (though men through the years have, of course, fallen in love with prostitutes)—is apparently a potent draw. Sex can be enjoyed (or not) on its own; the only thing the guy owes is his money. No guilt (at least not toward the prostitute) and no further obligations.

So guys visit prostitutes while they're on the road, or while they're at work (New York City cops have been busted for frequenting whorehouses within their precincts while supposedly on "active" duty). What is not accomplished on the street or in (sometimes state-sanctioned) red-light districts is supplied the world over by "escort services." Massage parlors provide a not very cryptic cover for otherwise frowned-upon prostitution. They are everywhere, and if some of them are no more than what they profess to be, the "no hand jobs" notice on a massage parlor ad seen recently in a tourist newspaper in Shanghai is otherwise the exception once again proving the rule. And if sex clubs like New York's Plato's Retreat are at the moment dormant in the United States, others like them will pop up again as soon as the moral climate and economic conditions again become conducive to such ventures.

Pornography in the United States alone is a multibillion-dollar industry. Peep shows, sex shops, nude dancing, even pornographic bakeries, abound.[11] Evolutionary psychologists will insist that much of this sex business, like rape, is an epiphenomenon of the *real* deal, which is, of course, having sex to make babies. Conversely, feminists are just as likely to maintain that pornography is misogyny, representing often violent exploitation of women by men. A sex shop owner, however, will say that sex sells and that he (or she) is in business to make a living and is in any case not getting rich. Just trying to eke out an existence, pay the rent, stay married, and, yes, possibly, as part of the picture, have some kids. The customers who buy the dildos, vibrators, and graphic video tapes will most likely say—when cornered—that what they do to take their pleasure in the privacy of their own homes is no one else's business. Some will turn out to have kids; others won't. Sex goes on, all the time, independently of the reproductive imperative, and some forms cost money for the practitioners.

Sex is by far the biggest economic item on the Internet, encompassing pornography, sales of sexual nostrums, penis and breast enlargers, relief for the erectily challenged—as well as direct personal connections, which in a few cases have led to tragedy. The Internet, as we already noted, is a nonstop purveyor of all things sexual—testimony not only to the power of the Internet in modern life but also to humanity's seemingly unquenchable thirst for pornography, sex toys, and sex aids.

So much for the obvious, direct connections between sex and the marketplace. But anyone who has ever perused a Victoria's Secret catalog for the pure titillation of it knows, too, that soft-core images of scantily clad women (and guys in other catalogs) is a powerful marketing strategy for clothes, jewelry, perfume, cars—any and all objects associated with the good life—with power. The message is "Buy this lingerie, this necklace, or this car, and you will feel great and have all the sex you are looking for." It's the power message, put to use to further the economic goals of those who want to sell you their material wares.

The connections between sex and economics run deep—and to the near exclusion of the production of babies. Neither the man nor the prostitute he visits wants a child as the outcome of their encounter. The woman buying a necklace to enhance her beauty and sense of self-worth is not necessarily thinking that this necklace is going to help her get pregnant. The guy buying the sexy car might think it will help him pick up babes, but as a rule he isn't thinking that the car will help him make babies.

If anyone's "fitness" (propensity to produce and care for babies) is enhanced by the sex industry, it is the people who are making the money off of it. And though this includes prostitutes, all too often they are sharing their fees with pimps or are, worse, slaves forced to have sex simply to eke out a bare, bleak existence. Mostly it is the entrepreneurs who make the money: the madams of the glitzier bordellos, the massage parlor and sex shop owners, and the pornography publishers.

And here, with those making (sometimes) handsome profits from the sex industry, there is indeed a connection between economic success and, at times, the production of offspring. A recent television documentary profiled a mother/son team in Los Angeles who specialized in purveying pornography centering on anal sex. They were shown making a good living (though being occasionally hassled by the powers that be). I have no idea whether the mother was able to have her son in part because of income derived from some earlier foray into pornography (more likely she was a relative newcomer to the game, just helping sonny boy out in his business venture).

But there can be little doubt that pornographers and other purveyors of sex in the marketplace use their incomes the way people everywhere, in all walks of life, use their incomes: to provide food, clothing, and shelter to themselves, to their spouse (should there be one), and to children (should they have them). The triangle of sex, economics, and reproduction is intact among the people who run the sex businesses. It is the *customers* who are buying their products who are committed to a 100 percent disconnect between sex and baby making as they pay their money to take their pleasures.

So sex is pursued on its own, for its own intrinsic, private pleasures. Sex is intimately linked to human self-esteem—to feelings of power (or lack thereof). And sex and money go hand in hand. Babies are incidental to all this. Humans have sex for lots of reasons, only some of which have to do with making babies and thereby passing along their genes.

# TEN

# Sex, Economics, and Babies in Social Systems

E very year, one or another of the supermarket tabloids designates the "sexiest man of the year," usually some good-looking actor who has recently been doing well at the box office. But most of the fascination is reserved for models and actresses in low-cut dresses: when we aren't reading about what a mess they've made of their love life, or how they're hooked on drugs, we're being told how beautiful they are—and how endlessly fascinating they are to their string of lovers and to all of their male fans who fantasize about them. The actors and actresses don't mind, of course, because sex sells and because what they are really after is promotional hype for their careers. Sex helping to make money, once again.

A couple of questions: Will these beautiful and sexy people have left more babies behind them once they finally shuffle off the stage of life than the average John and Jane Doe? Do wealthy people, regardless of their walk of life (or whether they earned their money or simply came into it the old-fashioned way) have more kids than, say, blue-collar workers? By and large, the answer is a resounding no: though I have been unable to marshal credible statistics, it seems a safe bet that actors and actresses end up having roughly the same number of kids as your average middle-class wage earner. Maybe fewer, given the self-absorption that seems to come with a life in show biz.

Same for rich folks—though ardent sociobiologists have written papers claiming otherwise. In fact, there is a deep disconnect in most societies (I'll mention exceptions below) between economic success and average number of children. That is crucial, simply because natural selection boils down to differential reproductive success based on some criterion—one attribute or another that some individuals in a population have more of, or better versions of, than others—conferring an ability to thrive, and to reproduce, that is better than that of less fortunately endowed individuals.

Bird wings may be used for display to attract a mate, but are not primary sexual organs used to have sex.[1] They evolved for flight—flight to get around to find food, flight in some instances for migration, and, yes, flight to find mates. Though wing size and shape vary a bit within each bird species, presumably natural selection honed the particular average wing configuration as optimal for the behaviors and physiologies of the birds within that species. Economic success begets reproductive success.

So, too, the peacock's tail, flaunted before the peahens and used as part of the entire ritual peacock mating display, figures directly in the reproductive success of the peacock. This is in keeping with Darwin's notion of sexual selection, where sheer excellence in sexual performance can, like more mundane relative economic success, lead to the production of more progeny than among less sexually successful rival male peacocks. In the world of animals and plants, there does seem to be a statistical (on the average), yet direct, linear relation between both relative economic success and out-and-out mating success—and the number of progeny any particular individual is likely to leave behind.

But the correlation is far weaker with humans. In keeping with the disconnect between sex and reproduction (recall that no other species, save perhaps our close relatives the bonobos and also common chimps, seem to share elements of this disconnect with us) in human beings, there appears to be a much more tenuous relationship between economic success and sexual prowess, on the one hand, and the relative numbers of babies produced, on the other.

This is not to say that natural selection no longer occurs in humans: though it takes place within the context of local populations living as parts of local ecosystems (a condition that is no longer true of most humans, and hasn't been since agriculture was invented starting some ten thousand years ago), natural selection can still be found working on humans. An example of this is the already encountered delicate balance between death by malaria versus death by sickle-cell anemia—a balance created because people with only one copy of the form of the gene (allele) that causes sickle-cell anemia do not develop the disease *and* are less apt to contract malaria than those who lack the sickle-cell allele entirely.

But this general disconnect between economic success, and even sexual prowess, and the production of babies means that natural selection has in most respects been greatly relaxed in modern human populations. In its stead lie cultural traditions—learned ways of being, doing, and knowing that are handed down through teaching and example, rather than inherited through genes. And though cultural traditions can change ("evolve") over time, and though a rough form of "selection" may be going on (as when transistors replaced vacuum tubes in electronics),[2] even in unambiguous cases where everyone agrees that some new invention is a hands-down improvement over the old ways of doing things, the idea does not spread because it increases the "fitness" of the inventor or of those who adopt the idea first. They might make more money, but the success of the idea—whether or not it will spread and take hold—does not depend on whether or not the inventor and early users thereby make more children. Why should it? Their genes have nothing to do with the cultural information being selected. Cultural evolution, despite its analogies with biological evolution, is a different cup of tea, simply because the mode of transmission of information in the two domains is vastly different.

So people have sex for pleasure. They have sex to make babies—sometimes on purpose, sometimes vaguely on purpose, and often by accident. They have sex, or get other people to have sex, to make money. Humans *sell* sex. Human sex is so far removed from the "traditional" role of simply being a means of reproduction that most of the old rules governing animal existence seem to apply only tangentially, if at all, to humans.

That is why evolutionary psychologists like to take the Pleistocene cop-out. Their mantra is often "We're not saying this or that behavior is adaptive *now*: but it clearly did evolve, sometime back there in our evolutionary past." That's how they get around all these blatant disconnects between sex and reproduction—and the equally obvious importance of culturally inculcated and transmitted human behavior patterns.

If the individual sexual and economic behaviors of human beings to a great extent are no longer directly connected to reproductive success, when we turn our gaze to larger-scale, social patterns of sex, economics, and reproduction, the picture is instantly much more complex and even more difficult to link to simple evolutionary biological percepts. The picture also turns considerably darker, for we find ourselves confronted with subjects like rape, sexual slavery and prostitution, differential spread of horrific diseases like AIDS, contraception, and human population growth—all closely connected to equally horrible topics like warfare, poverty, and famine.

These are, at the same time, all intensely personal topics. It is individuals who lead economic and sexual lives. It is individuals who become prostitutes and slaves, individuals who contract AIDS, individuals who rape and are raped, individuals who go on sex tours, individuals who make babies in varying numbers (or don't, whether by choice or otherwise).

It is equally the case, however, that it is entire social groups—not racial groups, but societies with different traditions and especially different (economically defined) classes within those societies—that produce the patterns of differential baby production and incidence of disease. It is when we step up from the behavior of individuals to examine the social patterns of sexual behavior and baby production that we see how simplistic the homilies of gene-centered evolutionary biology are. Often they are not just simplistic—but entirely wrong.

To be sure, cultural patterns of infanticide (and other means of controlling the production of babies), bride price, and other cultural institutions governing marriage and sexual behavior can be—and often are—interpreted according to the canons of evolutionary biology, sociobiology, and evolutionary psychology. Not all such interpretations are necessarily ipso facto wrong. Nevertheless, large-scale patterns at the level of class and society often reveal just the antithesis of the evolutionary psychological credo that human behavior is adaptive and revolves around the maximization of the spread of genes. Here we find example after example where, at the social level, maximization of reproductive success runs absolutely counter to the maxim that the successful spread of genes is what underlies both the evolutionary past and the contemporary expression of human behavior.

A stark case in point is rape.

## RAPE

Rape—physically forcing an unwilling person to have sex—covers a host of different sorts of events. Though women have been known to rape men, and both sexes perpetrate homosexual rape, usually "rape" refers to a male's forcing a woman to have sex against her will. In American society, in addition to violent encounters in alleyways, or forced entry into homes as a prelude to forced entry of the woman, we have such things as "date rape," rape of spouse (even when consensual sex is part of the marital mix), "gang-bangs" in bars—events in which the exculpatory plea very often consists of "she was looking for it." No one (usually) disputes that the dark-street rape is a violent crime, but in some other cases there is dispute: rape is a crime in which the victim is frequently blamed.

Boys will be boys. Though some evolutionary psychologists have come out strongly against this interpretation, others have insisted that rape is an evolutionary adaptation.[3] Maybe it isn't now, but certainly in the past raping was a good strategy to spread genes—or so goes the thesis. As a holdover from those days, men "need" to slip those bonds of fidelity with the nuclear mate, ramble, and spread their genes—and if they can't find willing partners, it is now ensconced in their genes to get the job done by force.

As Frans de Waal points out (see note 3), for this to be a strategy that succeeds—for rape to be a true evolutionary adaptation—it would have to be established both that the majority of men routinely commit rape and that rape habitually results in the birth of live, bouncing babies who will in due course proceed (whether by rape or more socially acceptable means) to pass on daddy's genes.

This is, of course, not the case. However much the argument that boys will be boys has at times prevailed in courts of law, rape is generally seen (even in American society, where apparently rape is more common than in most other industrialized nations) to be a crime of violence—one that ranks in seriousness just short of murder. You can even devise an evolutionary scenario why this must be so: rape involves someone else's daughter, someone else's mother, someone else's wife. Males, it might readily be imagined, see no advantage of having another man's uninvited genes added to the mix of family patrimony.

Most rapes do not result in pregnancy. Indeed, many apparently go unconsummated, since rapists commonly can't achieve or maintain an erection. But rape does, probably more often than not, result in ejaculation of sperm into the vaginal tract, and a certain number of those instances result in pregnancy. What then? What happens to those babies? Many fetuses conceived in rape are, these days, aborted. And even when the pregnancy is carried to term, the traditional (and socially sanctioned) course of action is to put the child up for immediate adoption. If that means the rapist has succeeded in spreading his genes, it surely means, too, that the prospects for a well-adjusted, "happy" life of the adopted orphan are far less assured than those of a child born of willing parents. The probability that that child will himself or herself go on to spread daddy's genes still farther in some fashion is arguably diminished by the circumstances of upbringing.

But male rapists attack nonfertile girls; they attack older, postmenopausal women. The "object" of rapists' attacks is by no means invariably drawn from the ranks of ripe, voluptuous young women—the best choice as a potential "baby machine" and, in any case, the stereotypically most sexually desired class of females in every society.

So, if by their very actions rapists don't seem to be out to make babies, what *are* they after? The short answer is that rape is an act of violence, but one of manifestly *sexual* violence. It is not just some guy beating up a woman; it is a guy sexually violating a woman—no less violent for all that—but the violence is overtly sexual.[4] What can that mean? Only one thing: men have feelings of power so tightly wrapped up in their sexuality that rape is an expression, in some deep sense, about power—sexual power.

More often, rape is the expression of the lack thereof. Even when impotence is not the final outcome, as psychologists and social workers have known for years, rape is basically an expression of rage—rage ostensibly against women (victims are usually anonymous), but rage born of a deep sense of powerlessness.

Sexual crimes are not performed by run-of-the-mill people who, whatever their problems (and everybody has problems), still function as "normal" members of society. If there is an evolutionary basis for human social behavior (and there certainly is), it is for social cooperation, for comporting oneself more or less within the dictates of the social group into which one is born. The old Darwinian conundrum of altruism—how to explain altruistic, socially cooperative behavior in a competitive, dog-eat-dog world where each individual is out for itself, for resources, for mates—is simply answered: in our social species,

*Homo sapiens*, where degree of genetic relatedness by no means explains why people "get along" with each other at school, at work, anywhere, the adapted way is the cooperative way.

Social deviance is *not* adaptive. Rape is socially deviant behavior—big time. So those who would wish to paint rape as adaptive fall back on the Pleistocene cop-out, saying, "Well, sure, rape isn't adaptive now. Never meant to suggest it was! But way back when, among our ancestors romping around on the African savannas, it was!" And given the lack of a fossil record of behavior that goes much beyond the production and use of stone tools, who is to say they are wrong?

But it is not true that there is no way to test the idea that rape was once an adaptive behavior in the human evolutionary past. We can circumscribe the problem. Though we cannot find out whether rape was common among australopithecines, or early hominids such as *Homo habilis, Homo ergaster,* and *Homo erectus,* and while we cannot be certain what earliest *Homo sapiens* was doing, we can do the next-best thing: we can look at the behavior of our closest primate relatives (the two "chimp" species in particular). And we can look at the documented behaviors of preagricultural, hunter-gatherer peoples as at least possible exemplars of how life was conducted by human ancestors on the African savannas.

Neither source holds out much hope for the hypothesis that rape is an evolutionary adaptation. Jane Goodall has described raiding parties conducted by male chimps at the Gombe National Park in Tanzania—raids to capture females that look for all the world like raids conducted by some modern human tribes. Both are evidently for the same purpose: to bring females from a neighboring group into their own—there to be integrated as members of the group and, of course, to participate in the economic, sexual, and reproductive life of the troop. Such behavior had long been considered a simple means of outcrossing—getting "new reproductive blood" into the community to stave off the deleterious effects of too much genetic inbreeding. Such raids—though they obviously do involve the forced removal of females from their natal bands, and include presumably obligatory sexual activity at least down the road in their new lives with their captors—are still a far cry from rape. Rapists are not looking for mates and mothers of their children.

The literature of sociobiology abounds with examples of "sneaky fuckers"—males who, when the dominant male(s) controlling reproductive rights for the nonce are caught napping, sometimes can approach and mate with a female. The degree to which such behavior leads to successful production of progeny is a matter of debate; in any case, sneaky fuckers aren't usually interpreted (by field biologists, both male and female) to be rapists so much as tolerated interlopers. Genuine forced entry into unwilling females seems to be a rare exception among our closest primate relatives.

What, then, of hunter-gatherers—those fast-disappearing groups of people who have never practiced agriculture, who, instead, live lives much like those of our ancestors? All their nutrients come from the hunting of animals and gathering of edible plants, meaning that their lives are wholly confined to the local ecosystem, and their numbers limited by its productivity commensurate with their cultural adaptations to wrest a living from what is often (to Western, industrialized eyes) rather harsh and unpromising circumstances. How widespread a practice is rape among hunter-gatherer peoples?

Not very. One searches the archives of the Human Relation Area Files,[5] if not quite in vain for references to rape in preagricultural societies, at least with very few "positive" results. It seems that the smaller the group (hunter-gatherer groups generally range in size between forty and seventy individuals, though some groups are larger), the greater the risk of social outrage when a woman is forced to have sex against her will.

There being no evidence that a strong history of rape pervades our evolutionary heritage, it seems much more sensible to conclude that the current commonness of rape is far more a reflection of the anomie of industrialized society than the vestige of the supposed ineluctable urge among males to spread their genes around as far as possible. Indeed, the whole idea strikes me as, well, stupid: it is an extreme example of an idea gone haywire. There is no reason to suppose that the biological drive to reproduce actually underlies all of behavior, and only by assuming a priori that it is so would anyone imagine that rape was a natural outgrowth of it.

But if the rape of one individual by another can in no rational way be interpreted in the simplistic evolutionary terms that have recently been mooted, rape at the social level really sends ideas that rape is about gene spreading right out the window.

## INSTITUTIONALIZED RAPE

Rape by individuals is social anathema. But we all know that there have been times and circumstances when rape has been condoned—even prescribed—for social purposes. The 1996 film *Calling the Ghosts: A Story about Rape, War and Women*, directed by Mandy Jacobsen and Karmen Jelincic, is a chilling portrayal of some of the earliest documented imprisonment, torture, and rape of Muslim and Croatian women from the Bosnian town of Prijedor. The story is told primarily by a woman lawyer then in her forties. At first reluctant to speak, she then comes to the conclusion that, despite the shame, she simply has to reveal what went on. Serbian army forces came to her town, at first rounding up the non-Serbian professionals and intellectuals. Though most were men, a number of women were also taken— removed to an unused factory, a camp known as Omarska.

The woman recalls a childhood when she and her friends were all blissfully unaware of who was a Serb, who a Muslim, and who a Croatian. Going back over the names of her girl- hood friends years later, she is able respectively to identify the Serbs, Croatians, and Muslims she knew. Some of her Serbian captors were people she knew by name—people who refused to acknowledge her once she became a political wartime detainee.

The ethnic strife attending the breakup of Yugoslavia was ostensibly just that: a revival of old ethnic hatred going back to the days when the Moslems first entered the Balkans and on into what is now eastern Europe. But what it was really about— as all wars have really been about since time immemorial—was land and resources. The Serbs literally raped and pillaged, a dif- ficult task given the heterogeneous makeup of the local popula- tions in Bosnia and Herzegovina. Serbian kids soon devised signals to let the troops know they were Serbs; otherwise there was no real way to tell at a glance.[6] But, using lists of names, they were able to cull the non-Serbs easily enough—and ship them off to the camp.

The men were tortured and, one by one, killed. The women were raped, at first the younger ones. But apparently all of them eventually, repeatedly. A colleague and a girlhood friend of the main reporter in the film, herself also a lawyer and, when abducted, a sitting judge, gets it exactly right when she says the rapes were all about humiliation. It was all about breaking the collective spirit of the non-Serbs. In an especially poignant pas- sage, she speaks wistfully of the surprise she and everyone else felt that this could happen in the modern era. But World War

II (not to mention more recent genocides in Indonesia and elsewhere) amply gives the lie to the supposition that we modern people are more civilized than our savage forebears and would never stoop to such unconscionable behavior.

"I kill you and I fuck your wife." That's the approach to "ethnic cleansing," localized genocide, and the total takeover of a region. Some women indeed became pregnant as a result of these mass, repeated rapes. But as the woman judge says, this was about the destruction of a people and (though she doesn't say this) not about spreading Serbian genes. Babies born of such circumstances may occasionally survive and thrive, but making babies is not what systematic incarceration, rape, and torture are all about. They're about genocide. The film is dedicated to the six women who never made it out of the camp.

Enough. It has to be said, however, that rape on this social level has always been part of humiliation and annihilation—of genocide. War may be "adaptive" if it conveys more resources, whether farmland in Bosnia-Herzegovina or the second-largest oil reserve in the world, in Iraq. The winners of wars might have more resources and their population might grow, making more babies conforming to the genes of the victors. But this has nothing to do with the competition among individuals within a single population for resources and, ultimately (at least according to hard-core evolutionary biology), for the spreading of genes through the production of offspring. War, though fought by individuals, is conducted at the level of organized social groups, and institutionalized rape in this setting is about the destruction of people, not the spreading of genes. It is a weapon of destruction, working in exactly the opposite direction of sexual intercourse for the purpose of making babies.

## SEXUAL SLAVERY: TOURS, ORGANIZED PROSTITUTION, AND THE ABUSE OF THE YOUNG

Get on the Internet, and do a Google search for "Sexual Slavery." You'll get lots of provocative ads, of course—from people all too eager to sign you up for a trip to Thailand, the Philippines, or Costa Rica. You'll also find all the ancillary promos for penis enlargers and all sorts of other sexual "aids."

But there will be nearly as many hits on articles detailing the nature and extent of the international sex trade—articles from human rights and similar websites, usually reprinted from the news media, aiming to expose the ubiquity of the international sex trade and the horror of its darker sides. This trade, like slavery itself, is a social phenomenon, more akin to institutionalized rape than to simple prostitution.

Not all prostitution is involuntary, and occasionally there are Internet articles as well about how prostitutes are organizing themselves in places like Paris and Canberra. Their major concern these days (a variation of the not unfamiliar plea that "a girl can't make an honest living anymore") is that the crackdown on the international sex trade is really an excuse to crack down on the local prostitutes, many of whom are keeping hearth and home together through their chosen line of work. Many are raising children: sex for sale in support of raising their kids. Hard to argue with that.

A case in point: the anthropologist Conrad Kottak's portrait of social change in a small fishing village on Brazil's northeast coast included a woman named Dora, considered a *rapariga* (whore) by her fellow townspeople.[7] Dora drew the contempt of her neighbors, not for having a child out of wedlock, but rather for apparently trying to support herself and her family in the absence of a working father. She moved back to her home village with her children after the failure of two liaisons and made ends meet in part by seasonal work tending house and cooking for the visiting anthropologists from the United States, who in the early sixties were showing up regularly three months a year.

Dora was by no means the only single mother in that village of 159 households (as of 1962). And not all earned the appellation of *rapariga*. But, as Kottak points out, there were damn few ways single mothers could raise enough cash to keep even their modest households intact. The main economy for the residents centered on the fishing industry; most of the men fished almost daily (except during storms—the "summer" field season for northern university people was the austral winter). Less obvious aspects of the economy included absentee ownership of property, especially the extensive stands of coconut trees that ran all along the barrier beach on which the village was built.[8] While some of the locals tended, guarded, and harvested the coconuts for the owners, the presence of the trees had little impact on the daily economic lives of the peasants who lived among them—until Arembepe became a tourist destination in the 1980s, the trees then played a large part of the picaresque draw. Even Mick and Bianca Jagger went there.

But in the sixties life was spare. There were electric lights only on two feast days. Single mothers occupied, as usual, the lowest rungs of the economic ladder and, *rapariga* or not, of the social order. According to Kottak, six of the seven poorest households in 1964 were run by single women. Some of the women were able to eke out an existence by offering goods for sale—mini-marts for fruit, tobacco, or, in several cases, various tropical fruit-infused alcoholic drinks. Some ran little bars for after-hours drinking—and perhaps other things. But the social pressure for women to do the right thing—meaning live with a man, keep his house, bear and raise his children—was enormous. Such expected behavior was also the most efficient as well as the traditional means of economic support for a woman in that peasant fishing village in northeastern Brazil. Those who could not or would not conform automatically had a tough time. Some of them became *raparigas*. According to Kottak, as the sixties gave way to the seventies, Dora if anything more openly embraced the life she had been accused of leading all along. Smart, vivacious, earthy, and hardworking, she did what she had to do to keep herself and her kids (several of whom died during childhood) going.

### Kidnapping and Sexual Slavery.

With the rise of the Internet, it has become increasingly difficult to distinguish voluntary prostitution from the kidnapping and unwilling transportation of persons over state lines, to be kept as virtual sex slaves in a foreign country. A recent film documenting the trade in young girls from Riga, Latvia, to Copenhagen and other European destinations makes it clear that some of the girls left home in search of work and that not all of them were entirely uninformed about what that work might be. Yet many of them end up living as virtual sexual slaves, and one of them is actually murdered in the course of her nightly duties. Yet the film also shows one of the Latvian girls on a pay phone, and we are told her conversation (which is not translated) is about lining up still more work, rather than about going home. Even in the throes of an amazingly economically robust international sex trade, the human condition is so variegated that the entire spectrum between enforced, involuntary servitude and free will seems to be there—sometimes difficult to tell apart in any single case and often at odds with one another.

Girls far "under age," as young as ten or eleven, are increasingly showing up on the streets of New York. Even if they are not victims smuggled in from foreign lands, their presence is a clear signal that a significant number of male customers seem to prefer children—or at least girls only in the earliest phases of physical maturation.

At the same time, it is obvious that children are much more helpless than adults, much more pliant, much more easily managed, especially if taken to a foreign land where they don't speak the language. In Asia, children are as often sold by the families as they are kidnapped. The kidnapping of children in India and other Asian nations has been documented well back into the previous century—and no doubt has gone on "forever." The kidnapping is never for ransom, nor does it seem to be for simply selling them to childless couples, since children are dirt cheap in places like India. Rather, it is nearly always to consign them to slavery to perform some kind of economic work for their owners. So in focusing on the kidnapping, selling, and use of children as sex slaves, I do not mean to ignore the fact that kids are commodities that in general can be—and are being—bought and sold for their economic labor as slaves.

### The Economic Value of Children.

It is worth noting here that child labor or slavery is yet another, extremely important link between elements of the three-cornered human triangle of sex, reproduction, and economics. Sex and economics are closely tied together in a way that for the most part has nothing to do, except tangentially or accidentally, with the production of children. But the fruits of the reproductive process—kids—are directly tied to economics when they are forced to work, often long hours in unhealthy conditions for little or no pay. Using kids as sex slaves connects the dots between all three elements of the human triangle, and accounts of what happens to many of these kids make Dickensian tales of sweatshops—or even stories of out-of-the-country Nike sneaker or Kathy Lee garment production—pale by comparison. Kids have always been used in the economic support of families—going back beyond the industrial revolution, probably all the way back to the invention of agriculture and the origin of settled existence: hunter-gatherer children play, learn, and gradually take on the economic chores that they will pursue throughout their lives. Economic exploitation of children runs the entire gamut from simple daily chores to selling them as slaves.

Nasty as it might be to think about, kids are not just the economic sinks that parents need to feed, clothe, shelter, educate, take to ballet lessons, and otherwise, in middle-class America at any rate, throw tens, even hundreds of thousands of dollars at before they are finally out of the house "for good" (and even then . . . ). The situations where the vector is reversed—and children are bringing in money vital to the support of their family or are simply sold as slaves for a lump sum—once again raises the issue of the connection between family (i.e., parental) financial well-being and production of children. For just as the inability to care sufficiently well for children has historically been a leading cause of infanticide, can it also be that children are deliberately produced as little working minions—in a pattern that truly does (in the opposite way that infanticide does) have direct effects on the "fitness" of parents?

The answer, of course, is yes. For example, the anthropologist Marvin Harris long ago pointed out that welfare policies that paid mothers living without an adult male in the household a certain amount of money per child had the unintended effect of maximizing single (female) households, and also encouraged the regularized, continual production of new kids as each one represented a positive change in income! No question that the use of children as economic producers (or the realization that they represent an insupportable economic drain) has led to adult reproductive behaviors that have either enhanced or depressed their "fitness"—meaning the number of children they have produced. That the fitness is often (though not invariably) increased among poorer people, and deliberately depressed among members of higher socioeconomic classes, is all too obvious—and a topic considered further below.

For the most part, however, little in the way of fitness (i.e., baby production) enters directly into the sex business, except tangentially via the profits of the entrepreneurs actually running the show. No one knows for sure how many kids (under, say, eighteen years of age) are "sex workers" or just how many sex workers—willing or not—are in the business. Some recent "guesstimates" yield numbers like four million women caught up in the international slave trade, some fifty thousand in the United States alone. It is also claimed that the sex trade in general is a $7 billion-a-year business—one segment of the global economy, at least, that is not currently on the ropes.

For many women sex workers, contraception becomes a major issue. The early jazz pioneer Jelly Roll Morton says that he made himself a "can rusher" in a New Orleans bordello just to learn the "2:19 Blues" and other songs by Mamie Desdume (he also claims to have invented jazz, thus casting considerable doubt on his overall credibility). Bordello contraception consisted basically of douching (often with coke bottles and towels), flushing out the sperm as much to clean up for the next John as to avoid pregnancy.

**Sex Workers, AIDS, and Contraception.**

The explosion of varieties of contraceptive devices—chemicals (however dangerous some of them have proven to be in their long-term disruption of hormonal physiology) as well as more sophisticated physical devices—have made the can rusher obsolete and the task of pregnancy avoidance in prostitution a lot easier. Beyond the rhythm method, beyond condoms, diaphragms, the pill, and RU-486, there are now implants and hormones, as well as a greatly simplified surgical procedure—all directed at women, and all aimed at preventing pregnancy as an unwanted outcome of sexual intercourse. A pill for men that depresses sperm counts seems destined to hit the market soon—a far more appealing alternative to vasectomy, which for most men strikes rather too close to the seat of power.

Of these various techniques, only condoms are at all effective against the devastation of sexually transmitted diseases. Of all of these ancient scourges, nothing compares with AIDS on the modern landscape. Since AIDS is closely allied with viruses found in African vervet ("green") monkeys, the scenario that the current pandemic of AIDS is a sort of "revenge of the rainforest" event occasioned by humanity's utter disruption of the tropical rainforest (hence a sort of forced closeness, including eating "jungle meat" and even possible sexual encounters, with wildlife), while by no means definitely demonstrated yet, remains plausible. Right-wing Christian ministers ranted on that AIDS was God's retribution for the unclean acts of homosexuality, until political pressures from both left and right recently forced them to drop that rhetoric. Now it is clear that AIDS is by no means strictly a white male homosexual affliction,[9] or spread to women largely through intravenous drug use by means of shared needles. Nearly half of the estimated 38.6 million people currently infected with HIV are now known to be women. Most women contract AIDS through sex with men.

Sex workers are especially vulnerable. In Sub-Saharan Africa, where the situation is bleakest, AIDS still spreads like wildfire, especially along trucking roads. There are stories of educated women—women working in the healthcare industry—who supplement their incomes to support their children as sex workers by night. Even these knowledgeable women routinely forgo the protections they know they should be using, because their customers still demand only unprotected sex.

Transcending the personal tragedy that AIDS visits on its victims and their immediate families, the spread of HIV is beginning to cripple the often already fragile economies of the Third World—a dark connection between sex and economics that possibly is something new under the human sun. Consider the famine that stalks much of southeastern Africa today. Drought may have triggered the decline in agricultural production, but equally important is the drastic loss of agricultural workers, who are either sick and dying of AIDS or staying home to care for afflicted family members.[10]

Nothing, to my mind at least, could be more poignant than the roller-coaster ride of the economy of Botswana, a landlocked nation of the Kalahari about the size of France that lies just to the west of Zimbabwe and Mozambique, two of the countries suffering famine at least in part because of the rampant spread of HIV. Botswana is an especially sad case for several reasons. The first is largely symbolic, but important nonetheless: though most of the country consists of arid grasslands (the Kalahari "Desert," which is technically not dry enough to qualify for the term; its grasslands are used for grazing by native wildlife as well as imported cattle), the Okavango Swamp in the northern reaches of the country is a garden of Eden that is one of the last nearly pure vestiges of primordial African wildlife. The ecosystem of the delta closely mirrors the ambient conditions of eastern Africa two or three million years ago—the environment where much of early human evolution took place, now frozen in the dry and otherwise barren-seeming outcrops of Olduvai Gorge and other famous localities running along the East African Rift system. It is the last best vestige of the environment that produced us, giving us insight into the very conditions in which our bodies—and behaviors—were shaped by evolution. Any downturn in Botswana's overall economic health is a serious threat to maintenance of this system, already under siege from drought and incessant attempts to throw open the green grasslands to cattle grazing, and to siphon the water off for urban human use.

Botswana had stopped being the British Bechuanaland Protectorate when it got its independence from the United Kingdom in 1966. When it became Botswana, its 330,000 inhabitants found themselves living in one of the poorest nations in Africa—and thus one of the poorest in the world. All that miraculously changed when an exploration geologist from the South African diamond conglomerate DeBeers discovered the source of the diamonds that had been desultorily recovered from the bottom sediments of the Limpopo River over the years. As a result, the Orapa mine (and, later, another source) was opened up in 1967, and the people of Botswana got to share in the new-found wealth as few Third World countries have ever managed to enjoy when the world's entrepreneurs march in and take whatever they want. DeBeers splits the proceeds of Botswana diamond production 75 percent to 25 percent, the lion's share, for a change, going to the Botswana people.

Instantly, per capita income rose markedly in Botswana. Young people were going to school both at home and across the borders. Many of them were training in the professions. Botswana's future looked bright as recently as the mid-1990s.

But storm clouds were already gathering, and today Botswana's 1.6 million inhabitants once again live in a desperately poor country. Their death rate is the highest in the world—24.5 per 1,000 population.[11] True, drought has attacked Botswana even as it has hit its neighbors. A cattle disease came over the border from Namibia some years ago, also depressing the cattle economy, which has been otherwise shored up by purchases from the European Economic Union. But the real story, the story of Botswana's rags-to-riches-to-rags tragedy is AIDS, spread along the truck route connecting eastern Africa with South Africa—a route that goes right along the eastern side of Botswana. AIDS is the reason why Botswana has the highest death rate in the world, and why its economy has all but collapsed, because all the diamond money in the world (and the diamond industry itself has been recently depressed, along with nearly every other segment of the global economy) cannot in and of itself keep a country's inhabitants healthy and productive. If two-thirds of the American economy depends on the spending of individual households, think how a rural agrarian economy like Botswana's must suffer when the AIDS infection rate soars.

The same story is being told around much of the rest of Africa; it is also the emerging story in Latin America, Indonesia, China, and India. It is a story in the former Soviet Union and, to an extent that we don't like to admit, even in the United States. HIV and the other, older, more entrenched, but nowadays seemingly less devastating, sexually transmitted diseases combine to form both a personal nightmare and a devastating force against the economies not just of individuals and their families but of entire nation-states.

The link between sex and economics on the social level looks darker and darker: for every happy hooker, for every person who thrives by selling sex in some guise or another, there are mounting statistics showing us that the disconnect between sex and reproduction—the move that made sex an economic commodity independent of the production of children, the move that has shaped much of what humans are and how they behave—has had unexpected consequences.

That many women become sex workers to feed their children is not only understandable and even laudable but also a move that can actually be interpreted as increasing her "fitness" in terms sociobiologists and evolutionary psychologists so dearly love. Nonetheless, however "natural," given the decoupling of human sex from reproduction (sex workers are not looking to make babies when they go to work), the personal health and socioeconomic consequences when a disease like AIDS enters the picture are anything but what Herbert Spencer meant when he introduced the phrase "survival of the fittest." The picture becomes even more complex when we turn our gaze to the connections between economic and reproductive success within entire groups of people.

## Human "Fitness"

Several factors make human fitness, defined by geneticists as the propensity to make children, rather different from, say, that of redwood trees or slugs—or virtually all other forms of life save those of our ancestral species and closest living ape relatives. There is, of course, the decoupling of sex from reproduction; crudely put, your average human figures to have had many more orgasms per year than your average lion, since female lions mate, always trying to make babies, only a few times a year should pregnancy not result the first time around (but they do couple multiple times during the several days that male and female are together). Though human couples in the developed world will read thermometers, tracking the optimum time for impregnation, most of the children born each year are not so deliberately sought.

But other forces at work in human life more directly affect a given person's reproductive output—his or her "fitness." They take the form of social rules, which are sometimes, but not always, written laws; ineluctable behaviors that arise from circumstances, such as what economic class you find yourself born into; social codes of behavior; moral strictures arising from religious or ethical principles; and, last, but far from least, free will—which when applied to matters reproductive is expressed as "choice" these days.

It is not always easy to tease apart these categories of external and internal influences on human reproductive behavior. For one thing, it does always boil down to one person's, or a couple's, deciding whether she/he/they want to reproduce, but the pressures to do so (from prospective grandparents, say) or not (the Chinese government, for example, if you are a Han and already have a child) are external to the prospective maters' personal desires. These pressures are relatively straightforward and can be followed or not depending on the consequences—presumably less drastic from the disappointed grandparents than from the Chinese government if the decision is to resist the pressure.

Other cases are less clear. Absent explicit laws, or individuals in your extended family telling you what to do, we still ask: Why do patterns of human reproductive behavior seem to repeat themselves over and over, most dramatically along class lines? How do the larger-scale "norms" of expected behavior—the things society expects you to do—filter down to influence an individual's behavior? For these are the kinds of issues lurking around the competing interpretations of abortion, infanticide (whether sex biased or not), and family size. Both abortion and infanticide, plus other determinants of family size, vary from country to country (reflecting different religious and cultural traditions), as well as between different economic classes within virtually any postagricultural (let alone postindustrialized) society.

How, in other words, does the social "infrastructure," so beloved by social anthropologists, actually work? How do norms and expectations regarding behaviors, often held in very diffuse, unarticulated ways, still manage to become translated into the actions of individuals? For it is especially in behaviors that are condoned in the doing but not in the mentioning—actions like infanticide—that these questions arise. The current practice (especially in parts of India) of detecting the sex of a child through ultrasound, then deciding to abort if the fetus is female, is deliberate—though the reasons for doing so are seldom understood or articulated by the parents who have decided to reject the child (more on this below). But denial is just as often a major player in the psychodynamics involving abortion and infanticide: although infanticide appears to have been rampant in eighteenth-century London, for example, it was almost always recorded as "accidental smothering" while the child was sleeping in bed, and its mother "accidentally" rolled over on it, tragically snuffing its life out.[12] Few people, and certainly not the mothers, admitted that the death was anything but accidental.

Less egregious issues, such as why poorer classes tend to produce more children per household than wealthier ones, are not so starkly immoral or illegal as abortion (at least to some eyes) and postpartum infanticide. Regardless of one's class status, it may be just as difficult for anyone to understand why poorer ("lower class") people tend to have more kids than richer ("upper middle" and "upper class") people as to understand patterns of infanticide—but at least when comparative production of children is the issue, there is no shame (indeed, perhaps even pride) in having more kids than the rich family living up the hill.

So how does the social "infrastructure" work? Because larger-scale social forces are at work, I prefer to think in hierarchical terms and to see individuals reflecting values and doing things that are perceived as the "norm." If causality were from below—meaning from inside—it would be coming from (possibly at least) the genes. The genes made me do it. But much of what we people do, even if we thought of it ourselves (or think we did), we do because it is a well-traveled path, something other people do. The psyches, sometimes even the consciences, of individuals pick up the social signals in the air. They know what is happening around them, and if the right circumstances arise, they will do what others before them have done—even if that means throwing an unwanted baby in a dumpster. Whether it is conscious or unconscious, whether people do what they do deliberately or simply because it vaguely seems to be the right thing to do at the moment, people behave in highly regularized, even predictable, ways. They do what they do over and over again. And some of the things that they do seem unconscionable—like killing their babies.

Infanticide would seem to be an antifitness bit of behavior par excellence. One thing is certain: there is no single infanticide gene that some people get and others don't. Infanticide (as we'll see) is not wholly inconsistent with the idea that people are out to maximize their genetic presence in the next generation. But there are other ways to view infanticide—for example, as a social population stabilizer and, at the very least, as an economic strategy.

## WHY INFANTICIDE?

Infanticide appalls everyone. It seems so inhuman. Yet infanticide is part of every known human culture. It is well documented not only in primates but in other animals as well— where sex ratios are biased despite the odds that, in most genetic systems, the expected outcome will be roughly half females, half males. Infanticide often, if not invariably, means that there are more boys than girls than expected, meaning that a preference for boys over girls is what underlies most infanticide.

The most recent census in China shows that there were 117 boys per 100 girls born in 2000, up from 114 boys for every 100 girls in 1990.[13] China's and India's selective female infanticide patterns are well documented and have attracted a lot of attention in recent years, even as women have begun to gain ground educationally and professionally (especially in India). But these patterns are hardly unique: infanticide rates of up to 50 percent have been documented by anthropologists in some hunter-gatherer cultures. Perhaps even harder to accept, medical investigators in industrialized nations have shown that at least some cases of "sudden infant death syndrome" (SIDS) are also, in reality, cases of infanticide.[14] A ratio of some 130 boys for every 100 girls has been documented in some bands in Amazonia (e.g., the Yanomami) and in late medieval England.[15] Like it or not, infanticide generally, and female infanticide in particular, pervades human society and has done so seemingly throughout history.

The question is why? In the by now familiar war between the realms of gene-centered evolutionary biology (including sociobiology and evolutionary psychology) and their allegedly benighted rival social scientists, infanticide provides a battle-ground for opposing interpretations. But cutting through the miasma of rhetoric, some interesting common points emerge—elements linked by the perception that economics has a lot to do with how much infanticide, and specifically how much female infanticide, is practiced when and where.

First the rhetoric: sociobiologists and evolutionary psychologists often complain that social scientists, especially anthropologists, don't want nasty facts about disparate cultures to be revealed, still less interpreted in evolutionary biological terms. For example, Sarah Blaffer Hrdy writes that the "devaluation of daughters was viewed [i.e., by social scientists circa the 1970s] as a purely cultural construct. It was assumed to be the outcome of free-floating minds spinning infinitely variable webs of meaning out of locally received traditions."[16] Though some cultural anthropologists (such as Marvin Harris) were not at all reluctant to admit biology into the mix of their analyses of human behavior back in the 1970s, most of them certainly considered cultural explanations to be more important than evolutionary biological causal explanations for patterns of human behavior. This is precisely why infanticide brings into such sharp relief the vastly differing interpretive views of human behavior.

A closer look shows that the differences between a gene-centered evolutionary approach and the more traditional take of anthropologists and other social scientists are not all that cut and dried. For example, anthropologists have long recognized that the reproductive output of a population is limited almost solely by the number of reproductively active females in the group. This is a simple biological fact—the very same one that led conservation biologists recently to urge the harvesting of does as well as of bucks (with their prized racks of antlers) during hunting season to control the burgeoning numbers of Virginia white-tailed deer in the United States. Harris even asserts that this simple fact—that population size is controlled by the number of females in a population—would hardly have escaped the notice of hunter-gatherers in the Pleistocene, people otherwise unschooled in modern biological knowledge.

Moreover, no one seems to disagree that infanticide—of infants of either sex—occasionally occurs when girls, ill equipped financially or emotionally to raise a child, disguise their pregnancies from others (and deny it to themselves), give birth unaided in secrecy, and abandon the infants, most of whom die. Mothers unwilling to acknowledge their babies' lives, mothers who are at some level unable to rear their children, do sometimes abandon them and, in better times, may well go on to have and nurture more children.

But why are there periods in history in some societies (but not, seemingly, in all) where abandonment of infants reaches epidemic proportions? William Langer, for example, has written, "In the 18th century it was not an uncommon spectacle to see the corpses of infants lying in the streets or on the dunghills of London and other large cities."[17] You don't necessarily have to argue that genes are behind it all to think that all group behaviors are the simple summation of the behaviors of individuals—and nothing more. You might (as indeed I think most evolutionary psychologists do) simply assume that when things are especially bad, the frequency of infanticide goes up strictly because there are more young women unable, for economic or emotional reasons, to face the prodigious demands of motherhood.

But there is clearly more to infanticide—especially differential female infanticide—than the tragedies experienced by individual women who cannot bear to produce a child at a given moment in their lives. One line of thinking arising from the gene-centered evolutionary perspective is an interesting—if unsurprising—gambit: it says that patterns of female infanticide, when taken in conjunction with historical traditions and current economic conditions, have everything to do with maximizing the survival of family lineages. Female infanticide, at least in certain times and places, is all about the survival and spread of genes.

For example, Hrdy turns to northern India to explore the notion that "there might be innate human predispositions that enhance[d] inclusive fitness and the long-term survival of family lines."[18] Reviewing the history, up to the present time, of infanticide in Rajasthan and Uttar Pradesh, she describes how sons of the elite classes, inheriting the family wealth, take as wives the daughters of families who are able to amass a considerable dowry.[19] The system leads to "hypergamy"—where women tend to marry men of higher social status—but "at the top of the hierarchy . . . hypergamy dooms daughters. There is no higher-ranking family for them to marry into." Hrdy relates how the British colonial government finally realized why so few women seemed to be around: few daughters born to the socially elite were allowed to survive. The farther down the social scale, the more fifty-fifty the sex ratios in these regions of India became.

Hrdy concludes that hypergamy, the practice of girls "marrying up" the socioeconomic scale is "a long-standing necessity for lineage survival. Nor can it be denied that decisions leading to it have genetic consequences." While she is right that there are "genetic consequences," it is all too obvious that female infanticide is not confined to elite classes in India or anywhere else. And what is inherited is wealth—land and valuables. There are always fewer elite—hence fewer of the genes of the elite—than there are intermediate and lowest classes. Rather like the case of the Florida scrub jays (chapter 6), where helping to raise later offspring of one's parents was widely cited as an open-and-shut case of "enhanced inclusive fitness," when what was actually being inherited was hard-to-come-by territory suitable to life as a scrub jay, reproductive patterns in the elite of northern India were about the survival of wealth—and enough genes to minister to it. Sure, "keeping it in the family" has a prima facie genetic corollary, but it is at least as likely that it is the survival of the wealth, rather than of the genes, that is at stake. The elite in every society hold their power through their wealth, not through their sheer numbers (of individuals, or of extended shared genetic patrimony). The relationship between economic and reproductive success in humans is almost always 180 degrees different from that in the rest of the living world, where economic success as a statistical rule leads, as a side effect (and all other things being equal), to the production of more offspring than does inability to cope with the economic exigencies of life. Darwinian natural selection at its core.

Clever (though flawed) as this scenario of infanticide, marriage patterns, and the survival of blood lines in northern India is, it doesn't begin to encompass the full gamut of infanticide—especially differential female infanticide—in all the varying ecological settings (hunter-gatherer versus postagricultural and postindustrial societies) and in social organizations ranging from bands and villages through the huge nation-states of today's world. If genes aren't driving everything, including infanticide, then we must consider alternative, cultural explanations. And it is simply not true that social scientists have been unable or (worse) unwilling to grapple with these issues.

Most biologists and social scientists would agree that infanticide, like many issues broached in this chapter, is so complex that no single theory (like survival of bloodlines) is likely to explain all its manifestations. Nonetheless, some daring anthropologists have taken up the challenge. I'll mention briefly here only the ideas of one of them—Marvin Harris, to whose inspiration this book is dedicated. This is not to argue that Harris is absolutely right and that gene-centered theorists are completely wrong. Rather, I want to illustrate the lines of argument and evidence that can be brought to bear on the explanation of social reproductive patterns such as infanticide; these ideas interweave some ineluctable biological, and potentially genetic, facts that determine what people do, with the higher influences ambient in the larger-scale social system—influences Harris referred to as infrastructure.

Harris sought to establish a scientific basis for the study of the diversity and history of human cultures—to lift anthropology from the descriptive cataloging of cultural diversity, to see instead repeated patterns in human history, and to interpret them in terms of the processes that caused people to behave in the ways they are observed to do. He was a "cultural materialist," believing that many of the mores and traditions of individual cultures reflect a behavioral adaptation to the specifics of the world that they found themselves living in. On issues of economics and reproduction, Harris formulated a neat equation between reproduction and ecological/economic production, especially the production of food, first by hunting and gathering and later by the controlled production of plant and animal life—"agriculture."

Harris, like many critics of Thomas Malthus, did not see

human population growth as inevitably reaching a crash-and-burn point, where the production of babies would overwhelm the capacity of society to feed all hungry mouths. He did acknowledge the simple biological fact that populations will grow (exponentially) in all sexually reproducing species if parents as a rule produce more than two offspring—if they do more, in other words, than merely replace themselves. This was the starting observation for Malthus's dark ruminations in 1799 on the future of European populations and, later, for the Darwin/Wallace Aha! of natural selection: that because populations of all animals and plants seem to be in some form of stable equilibrium, something must be keeping their numbers in check—and that only the most economically successful of them are likely to survive and reproduce.

But Harris, taking the longer view, saw the relation between baby production and productivity in human groups as, potentially at least, a two-way vector. Of course, populations will grow and quickly approach the "carrying capacity" of the system, be that system the scant animal life and tubers of the Kalahari or the potato fields of Ireland. And disasters do indeed happen, as the blight that struck the Irish potato crop a few decades after Malthus pondered the issue certainly shows. But when that carrying capacity is reached, humans have more options than the rest of the biological world: their numbers can be stabilized simply by the limits of what is available (as in all other living populations); or they can *choose* to limit their population numbers; or they can reinvent themselves—specifically the technologies with which they wrest a living from the earth. In a nutshell, Harris saw human history as periods of temporary population regulation/stasis according to the dictates of their mode of production. But eventually, perhaps even inevitably (and in rather Malthusian fashion), pressures would build up so that the best option was to "intensify" production. Each of this succession of "energy crises" led to an expanded carrying capacity, and thus to the survival of more and more humans.

But "energy crises" and the consequent intensification of production need not have happened that way, had humans had the efficient means of regulating birth that appeared on the scene only recently: condoms, diaphragms, the pill, RU-486, and all the more modern hormonal and other chemical means that are on the market today. Harris actually believed that, given the choice, people would limit their population size rather than go through the agonies and uncertainties of intensification. He held that, had the gamut of effective means of birth control been available all along (ironically, they are themselves an outgrowth of modern technology, one of the spurts of "intensification" that has led to population growth), human population would have stabilized long ago. In other words, he believed in free will and reproductive "choice" (in the more general sense), though he also thought that it was perhaps too late for these remedies to be put to effective use in staving off the rampant growth in human population currently engulfing the world. And he maintained that, for the greatest part of human history, the most common human form of birth control was infanticide.

Harris thought that infanticide in general, and female infanticide in particular, is a form of population control and is related to larger issues of group economics. For example, the Yanomami people of Amazonia, whose villages are commonly at war with one another, typically show the ratio of 130 boys for every 100 girls (or higher) during times of war, yet the ratio of adults is usually nearly 1 to 1. The balancing presumably occurs through the loss of males in warfare. Typical of Harris's reasoning is his conclusion, regarding the Yanomami, that "the peak in warfare results from pressure to maintain living standards by exploiting larger or more productive areas in competition with neighboring villages, while the peak in female infanticide comes from pressure to put a ceiling on the size of the village while maximizing combat efficiency."[20]

But female infanticide is by no means limited to situations where neighboring villages are at war. In fact, Harris asserted, female infanticide tends to occur whenever and wherever males are more valuable economically—such as in Eskimo groups, where food production relies almost solely on the labor of men. He went on to contend that, in less arduous ecological climes, differential female infanticide becomes less common, in direct proportion to the rising economic value of women.

There are, however, as many problems with this blanket cultural scenario of female infanticide as with the notion that it represents a strategy for maximizing the passing of genes of particular lineages along to the next generation. Eskimo women perform many economic tasks—though it is true that the actual gathering of food by women living in hunter-gatherer groups in the tropics is a more immediately direct contribution to "production" than the preparation of food and the countless additional, patently economic tasks all women everywhere perform on a daily basis. But women in China have for ages been strongly involved in primary food production. And reports from India make it clear that, despite recent improvements in the educational, political, and economic lives of women, differential production of boys over girls (increasingly through prenatal abortion) is, if anything, on the rise.

Once again, why? Is it all just "tradition," as some evolutionary psychologists are fond of accusing social scientists of saying? Is it, rather, a reflection of individual aberrant behaviors writ large (the dumpster scenario)? Is it because of the ineluctable urge to leave more copies of your genes to the next generation, and because males have a higher probability of doing so, especially in some forms of social organization where women tend to marry up and have a dowry attached to their marriage that poorer parents can ill afford? Is it, rather than the upward causation driven by genes or the behavior of individuals acting in isolation, a reflection of economic necessity, given the ecological situation people find themselves in? And, as such, could group-level factors (e.g., the population size of the village) matter more than the effects of differential infanticide on the preservation of individual "blood lines"?

Perhaps unsurprisingly, some of the more recent ideas on why female infanticide/differential abortion is especially common in India and China (and other Southeast Asian nations) draw on a mix of these explanations. Nearly everyone seems to agree that China's one-child policy (which is, in any case, often ignored or avoided through official dispensation, particularly in rural areas where families need workers for the fields) has, if anything, exacerbated the problem: if only one child is allowed, then it had better be a boy. Focus has recently shifted to the intensely patriarchal nature of most Asian cultures; not only is land and other wealth inherited by the sons, but so is the care and feeding of mother and father: once a woman marries (*if* she does), her responsibilities to her parents supposedly end pretty much right there. It is alleged that this heavily patriarchal system of inheritance and responsibility is far less common, though not unknown, elsewhere in the world.

So, moms and dads prefer sons over daughters, because that's their form of social security, ensuring their survival in old age. However much daughters may be involved in the workplace— in the fields, in the factories—it is the sons who, pretty much the world over, at least traditionally, are the main source of economic input into the household.

If this is the case, we have to ask whether it is the inheritance—the transfer of wealth to the next generation, along with the assumption that there are automatically "genetic consequences" to such a transfer—that is driving this system. It's those putative "genetic consequences," naturally enough, that most evolutionary psychologists are looking for.

Or is it more that parents look to their children for economic sustenance in their old age, especially in societies where there are no pension plans or government forms of "social security"? Though evolutionary psychologists say that humans are endlessly fascinated by the histories of families, be it their own or important families in history or even in the movies, there is depressingly little evidence that people care what happens to their own family more than an average of a generation and a half down the line—which in modern American society means seeing your grandchildren survive into their thirties. A generation and a half translates into a person's average life expectancy. People tend to care only about what they will see when they are alive—and, as a rule, are greatly interested in staying alive. People know that when they are dead, they have no control over, no stake in, what lies ahead. *Après moi, le déluge.*

Where female infanticide occurs, it does so because sons are valued—that much is obvious. That the value is in some sense economic is also clear. That it is an unconscious, socially inculcated custom either to regulate population size or to maximize the persistence of genes down through the generations may be true in some instances, but it is not an utterly convincing, blanket explanation of the phenomenon.

That infanticide generally and female infanticide in particular are socially accepted practices that have profound economic consequences seems likely. That those consequences are for the short term, the here-and-now benefit of those who are alive, is also likely. Keeping older people alive, long past their reproductive years, may nonetheless contribute to the welfare of any grandchildren that are around. For the most part, however, it is the parents who want to be alive, and to continue to enjoy the economic side of their lives.

In fact, short-term economic gain seems to underlie much of the human reproductive behavior down through the ages. Whatever else it might be about, infanticide hovers around the theme of saving money (can't afford either to raise the child or supply a dowry for a daughter), channeling wealth toward descendants (who use it in part to keep you in your old age), or biasing the production of able-bodied males to enter the economy. It is exacerbated by laws restricting the size of families. Historically, it may also have been a socially inculcated way to keep band size (as in hunter-gatherers) at an optimal, sustainable level.

It is far less certain that infanticide, especially differential female infanticide, is all (or even partly) about channeling genes of individual family lines to the next generation. Selective elimination of infants seems to have a strong, and very short-term, economic motive—one that is overcome by the time reproductive age kicks in and sex ratios come into balance.

## THE MAYOR OF SINGAPORE, OR, "BEGGARS BREED AND RICH MEN FEED"

Sometime back in the 1980s, the New York newspapers carried the story of the distress felt by the current mayor of Singapore.[21] Why is it, the mayor lamented to the world press, that all our best-educated, brightest, and most successful people seem to be having so few children—while our poorest people are having them with alarming regularity? We need more of the productive types to push the economy further. We don't need more of the sorts of economic drains on society that the poor represent with their ever-expanding numbers.

Singapore began offering incentives (substantial tax rebates) to better-educated, higher-income families to produce more kids—and, as direct cash incentives, to lower income, poorly educated individuals to produce fewer kids, including U.S. $10,000 to any woman under thirty who would agree to undergo sterilization. From 1984 to 1987, the gap in number of live births between the wealthy and poor dropped: the plan worked. But then the government decided that, in the currently booming economy, Singapore was not producing enough native sons and daughters for the needs of the labor force, and it changed the policy to across-the-board financial incentives to produce more children. The so-called dysgenic gap (disparity in reproductive rates between rich and poor) has long since reappeared, and the mayor's successors are back to square one: forced to confront the choice of having fewer, better-educated Singaporeans or a sufficiently large labor force to meet the demands of their own industry.

Back in the 1980s, the mayor made it clear that he considered Singapore's situation an anomaly that somehow ran counter to the natural scheme of things. After all, isn't the normal order of things, in this dog-eat-dog Darwinian world, that those who prosper most tend to leave behind more children—more copies of their genes? That's what the mayor thought—thus his bewilderment over Singapore's predicament: not only was it a short-term economic problem that needed solving; it was also strangely out of kilter with the way things naturally are and forever should be.

The mayor was not speaking in a rhetorical vacuum. His outlook already had a long history. According to Gail Collins,[22]

Teddy Roosevelt and other notables complained especially about the reproductive failure of 75 percent of the women "entering professions like college teaching, social work and library studies" in "soaring" numbers between 1890 and 1920. As we already noted, "survival of the fittest" was Herbert Spencer's, not Darwin's, phrase, and though Spencer himself was not perhaps the prime mover, "social Darwinism"—the application of Darwinian biological principles to the human condition—took root in the Victorian world soon after 1859. It might even be said that, as a rationale for the often ruthless tactics of Victorian capitalism on up through today's big business, the rationalizing of the competitive, devil-take-the-hindmost human behavior in the name of Darwinian biology probably did more for the social acceptance of the idea of evolution than did the revelation—repugnant to so many—that all life, including our own, is descended from a common ancestor in the dim recesses of geological time.[23]

One thing these budding early Darwinian interpreters of the human social condition did get right: for the principles to work in the biological world to produce evolutionary change, economic success has to be translated into reproductive success. That these rules blatantly do not apply the same way in human society as in the rest of the biotic world confused, worried, and angered many social critics. For what does mere wealth mean if the rabble at the gates will outbreed you and end up, in more ways than one, eating your lunch?

Is it really the case that in all stratified societies the poor reproduce at a faster rate than those better off? After all, wealth might be expected to be concentrated in the hands of the few. Given that assumption, the wealthy will indeed remain numerically small, even if they breed at the same rate as those with less money. Larger numbers of people will produce more children than smaller numbers, even when they are reproducing at the same rate.

So the question really is this: Do poor people on the average per capita make more babies than the rich? The answer—any way you look at it, from the details within a city (or city-state, like Singapore) to a global perspective—seems to be a resounding yes. More babies are born to families that tend to be less well off than those in the upper echelons of economic status— the wealthy. With exceptions, of course.

It's easiest to see this on the global scale, comparing rates of production of children with the overall wealth of a nation. The same countries with the highest "crude birthrates" and "fertility" rates are those with the lowest gross domestic product (GDP) and purchasing power per capita: countries such as Ethiopia, Congo, Liberia, and Niger.[24] On the global scale, it is the have-nots who tend to out-reproduce the haves.

Within a single country, the inverse relationship between wealth and fertility is often manifested along class lines. This happens especially where one ethnic group holds the political power often to the virtual exclusion of the other(s). Consider the modern state of Israel. According to one source, the "1948" Palestinians living in Israel circa 1995 made up some 19 percent of the population; their population (excluding immigrants) was growing at the rate of 3.1 percent, compared with a rate of 1.2 percent for that of the Jews (again excluding immigrants). The infant mortality rate (per 1,000 live births) was 9.7 percent for Palestinians and 4.9 percent for Jews. The average persons per household was 5.1 for Palestinians and 3.3 for Jews, and the average monthly income of salaried urban males was 2,494 NIS (new Israeli shekels) for Palestinians and 4,555 NIS for Jews— with roughly half that amount, in similar proportions, for salaried urban females. Another Internet source claims that Palestinian per capita GDP has fallen since 1993. This is by no means the whole story, but there is enough here to suggest that even in places where the underclass is not numerically dominant (as is the rule in most ethnically partitioned modern states, where political power rests in the hands of the numerically dominant as well as more prosperous), the families of the underclass have less money and more people than families of the dominant class. Nonetheless, if the numbers are to be trusted, it is the poorer people who are having—and raising—more babies than those who are in power and who, on the average, have more wealth.[25]

Despite the efforts of some sociobiologists to deny it, the same pattern persists in developed nations such as the United States. The U.S. Census Bureau's numbers on and analysis of the fertility of American women as of June 2000 reveal the same patterns: in eight categories of annual family income (ranging from "below $10,000" to "above $75,000"), the number of children born per 1,000 women was 1,394 for the poorest category and 1,066 for the wealthiest category.[26]

Sociobiologists occasionally take issue with what they regard as the anecdotal, largely mythical claim that poor people tend to outbreed richer people, looking instead for a positive correlation between economic and reproductive success. One attempt used the pages of *Forbes* magazine (and additional sources) to measure the reproductive output of four hundred of America's wealthiest males—and found that these people tend to have slightly higher numbers of children when compared with the general population.[27] Taking this study at face value, we certainly see that in some situations, powerful, wealthy men will have more children than the average guy on the street. Ramses II of Egypt supposedly had children in the thousands by his many wives, and ethnographers have encountered this pattern in many different tribal settings. On the other hand, the custom of seignorial right (allowing the landowner to bed a new bride among his peasantry on her wedding night) is more an instance of the tyranny of political rape—sending a message of absolute power and control to husband and wife—than an example of the rich guy getting all the chicks and making most of the babies. Even if it were true that rich guys (though not, evidently, wealthy women) on the average make at least as many babies as less well-off males in places like the modern United States, the power is concentrated in their wealth, not their genes. Lest there be any doubt about which is the more important of the two to be handed along to the next generation, consider the Forbes family.

So the old mayor of Singapore at least had his facts straight: the poor of Singapore were reproducing at a faster rate than the rich. On the face of it, it is thus definitely the case—on all sorts of geographic scales, from cities to the global population as a whole—that fertility is higher among poorer people than among the more affluent.

It is also the case that the trend can be reversed and that fertility rates are dropping in many parts of the world, slowing globally from rates projected only a few years ago. Not only is this good news for the overall well-being of Earth's ecosystems and the average quality of life of its human inhabitants; the recent decline in birthrates among the poor as well as the wealthy holds the key to understanding just what the relationship between economics and reproduction in human beings is all about.

## THE ECONOMIC SIGNIFICANCE OF CHILDREN —AGAIN

Who needs kids? What are children all about? Are they merely the vehicles by which our genes manage to propagate themselves?[28] Are they simply the little darlings (troublesome as they can be) that make their parents' emotional lives complete? Or does the existence of children mean something more to their parents (and society at large)—namely, their economic significance?

In the cosseted world of upper-middle-class American life it's perhaps true that a kid's main job is to figure out its parents, upon whom it utterly depends for its own continued survival. Kids are living purely economic, nonreproductive lives for at least their first decade and a half (for the most part), and growing, developing, and just leading their lives does amount to a kind of work—however nonproductive that work might be in terms of household economics.

So we tend to think of kids purely as economic drains: all the money that needs to be spent on birthday, Christmas/Hanukkah/Kwanzaa gifts, how they eat their parents out of house and home, and how outrageously expensive their high-tech sneakers are—sneakers that were made in Shanghai, at least in part by children.

Children do perform real work and have done so through the ages. If the job of boys in bands of San people living in the Kalahari is to learn how to shoot an arrow and stalk game, copying the adults while playing with their peers at a distance, before long they are big enough to join in the hunt for real. The girls learn from the mothers the tools and tricks of the trade of assembling thatched houses, finding tubers and berries, preparing meals—and are soon actually helping out in those chores.

Children everywhere, in nearly all conditions, are thus both drains on and sources of input into the household economy. In the rest of the biological world, a useful distinction can be drawn between species that produce relatively few offspring, and often are raised with a great deal of parental investment in time and energy, and species, even closely related ones, that produce a great many offspring, and invest little or nothing in seeing to it that the offspring survive. The thought is that producing relatively few well-nourished descendants is one way to see your genes make it into the next generation, but so is producing hundreds, even thousands or millions, of offspring, throwing them out into the cruel world, where many will die, but some will survive to adulthood and start the reproductive cycle all over again.[29]

We've already encountered some of this sort of thinking applied to humans: the life expectancy of poor children is usually much lower than that of more affluent children. Poorer people have fewer resources for their children—and anthropologists have reported cases where dire economic straits end up channeling proportionately more resources to the main wage earner (usually the father), followed by the mother, and then the children.[30] In all other species, a sizable spectrum of offspring produced would be unstable in most situations—as sheer numerical dominance (and the genes underlying it) will win out sooner rather than later.

Something about the human situation, however, is different from purely biological systems—something that allows a K-selected-like behavior among the more wealthy and favors a more r-selected-like pattern among the poorer classes, without one's giving way to the other. That something is the purely cultural way in which economic resources are generated, concentrated into the hands of the relatively few, jealously guarded and maintained, and then handed down.

Well-to-do people may complain about college costs, or the price of their daughter's BMW, but they pay the tuition and buy the car. The wealthier parents are (in the context of the society in which they are living), the more they can stand the "economic drain" side of raising children; indeed, like the Sopranos', their children's expensive clothes and education become something of a status symbol every bit as much as their own McMansions and luxury SUVs. The coddled kids will survive and, if not too emotionally disturbed, go on to lead similar lives and perhaps produce grandchildren.

The poor, meanwhile, with far less to go around for themselves and their kids, have *more*, rather than *fewer*, kids. And though part of this seemingly anomalous pattern might reflect the r-selected-like sense that, the more babies you make, the greater your chances that some will survive and thus your genes persist, there is more to it than that: it is precisely the children of poverty that have been forced to work as miniature adults—to enrich the family coffers or, far worse, to work as slaves for someone else's enrichment. It has become common knowledge, in recent years, that many of the Indian rugs so eagerly sought by affluent Westerners are produced by enslaved child laborers. Indeed, kidnapping of children in Asia and elsewhere (a practice that goes back at least a century, and probably millennia) for sale as economic (as well as sexual) slaves is rampant. Children are economic commodities, themselves being bought and sold.[31] And poor children are the most vulnerable of all.

Most child labor arguably benefits the family finances directly. The more hands the better; they can all be put to work. Beyond whatever direct economic worth young children may have for their poor families, the question is again what's in it for the parents to have so many children once the children are grown. As in the case of infanticide (where evolutionary psychologists tie themselves into knots trying to make the killing of offspring at least consistent with a viable parental strategy for perpetuating their genes), it is clear that the more kids you have when you are desperately poor, the more likely it is that some will survive to go on and make babies themselves—an r-selected way of passing the parental genes along.

The alternative is to focus not on the genes but on economic wealth. As we've seen, the latest thinking on patterns of differential infanticide in Asia hinges on which of the kids will end up caring for their parents in their old age. The argument is that in Asian systems, where care of the older generation falls to the sons, not the daughters, daughters become expendable (especially when there are limits imposed from above on the number of children a couple are allowed to have).

The same kind of explanation applies to the positive production of offspring—in poor families where there are no restrictions on family size, where the parents need someone to survive merely as their own form of social security. It's not about passing genes—or even necessarily money, land, and worldly goods—along to future generations; it's the far more short-sighted desire to have your postreproductive body, not your genes, be looked after.

That cultural patterns connecting economics with child production run so habitually counter to "normal" patterns in the biological world is reason enough to conclude that the rules governing human "fitness"—at least in post–agricultural revolution societies—no longer resemble those of the primordial biological state from which we ultimately sprang. As the Singapore saga clearly shows, input of financial resources into the system can, by sheer bribery, increase the reproductive output of the wealthy and at the same time depress the reproductive output of the poor.

## FITNESS, CLASS, AND HUMAN
## POPULATION GROWTH

But on a grander scale, accidents of economic history, as well as more recent consciously devised programs to increase education (including information about birth control), show that economics runs the human reproductive show—and does so in the opposite way economics and reproductive success are connected in the process of natural selection. The media today are filled with the sorts of concerns that caused Singapore to abandon its policy to reduce disparity of baby production between rich and poor in favor of just making more local members of the workforce. First Japan, then Italy, and now more and more countries (especially in Europe) have become alarmed over their plummeting fertility rates—down to as low as 1.2 (2.1 is the number needed to maintain a stable population number; the United States is currently at 2.0). "We cannot afford children," seems to be the general cry, while governments worry about workforce and who is going to pay for the enormous costs of retirement benefits when the current crop of baby boomers retires.

The global population has gone from a scant five or six (estimated) million humans at the dawn of the agricultural revolution only some ten thousand years ago, up to six billion plus now. Though growth was slow at first, there is no doubt that, once humans stepped outside the confines of the local ecosystem—where the numbers of any hunter-gatherer band are sharply limited by the natural production of edible plants and animals—the potential existed for unprecedented human population growth, provided productivity in excess of the world's human-inhabited local ecosystems could be established. This occurred initially by simple agriculture and animal husbandry and subsequently by a series of intensifications (to use Marvin Harris's imagery), as technology improved in successive waves ("revolutions"). By around 1800, there were a billion people, two billion by 1930, three billion by 1960, four billion by 1974, five billion by 1987, and six billion by 1999.

Let's assume that the destruction of fisheries, forests, fresh-water systems, and topsoil through overexploitation, pollution, and simple habitat conversion poses an eventual threat to human life or, at the very least, puts some upper limit on how far our numbers can continue to increase.[32] It would then seem reasonable for us to give some thought to how we might stabilize our numbers before the limit of the system harshly imposes its own ceiling on our numbers.

Of all the approaches to stabilizing human population size, the most popular until recently was to raise the GDP of a region, for economic development does in fact go hand and hand with a slowdown in reproductive rate. Just as families on the richer end of the spectrum in the United States tend to have fewer kids than people living in poverty (whether in cities or in rural regions), richer nations (as we've already seen) tend to have fewer kids. And increases in prosperity tend to depress birthrates even lower.

But, though some demographers and economists propose economic development of Third World nations as the best way to stem the tide of their still-soaring populations, lack of resources and education in such already overly stripped regions makes the prospect of significant economic development in many Third World (especially tropical) nations a forlorn hope—if not a downright cynical joke.

Some genuine promise for reducing fertility rates in Third World nations lies in a related, albeit alternative gambit: the education of women and continued integration of women into their country's economy. Onlookers are surprised that, despite the very real gains along these lines that they have made in places like India, women are if anything having more ultrasound tests and aborting their female babies. But they are also beginning to have fewer children.

None of these sweeping national and global patterns of connections between economics and baby making have much to do with the survival of individual bloodlines. True, it is individual men and women who get together and make babies. True, the desire to have children is not only socially inculcated; it represents a deep-seated biological urge that is over three billion years old. But the patterns of baby production that we see, and especially their relationship to the economic lives of people, are profoundly different from the equation "Economic success implies reproductive success" of natural selection in its original, most pristine (and still most valuable) form. This disconnect between human economic success and production of children underscores the stunning lack of caution that biologists at least as far back as E. O. Wilson have shown in interpreting human behavior in simplistic evolutionary biological terms. It is rules of culture imposed from above, rather than an irresistible drive to leave as many copies of one's own genes to posterity, that ultimately control the human reproductive roost—and the lion's share of the rest of human behavior to boot. And those cultural rules, such as China's one-child-per-couple law, often work directly counter to the expectations of selfish-gene theory.

Only if we were to maintain—against all evidence and reason—that there is a genetic cause for social stratification would there be any motive to persist in a gene-centered view on the major aspects of human life. Classes are not always ethnically pure by any means. But there is enough class stratification with strong overlap of racial identity, in the United States as elsewhere, that racist explanations for why some people are poor, others rich, persist to this day. Evolutionary biology has, pathetically enough, long played a role in such racist accounts—a role it shows little promise of abandoning. The roots and maintenance of the classes defined by poverty and wealth lie in the interplay between that wealth or its lack, on the one hand, and the reproductive strategies people the world over, regardless of ethnic identity, characteristically adopt, on the other. And those strategies are the opposite—as the mayor of Singapore trenchantly pointed out—of what would be expected under Darwinian rules.

# ELEVEN

# A Moral to the Story

I t is ironic that the assumption that genes rule human life is itself a purely cultural belief. True, it supposedly has the support of modern science (bad biology, in my book), but it's a cultural belief just the same. It is ironic because this deeply flawed belief is often used to bludgeon the notion that cultural ideas are important in human life!

There is no mistaking that ideas shape human life. The United States is a nation of laws not because God created us in His image and charged us with behaving in a moral manner befitting our origin (as seemingly believed by the leading creationist and legal scholar, Phillip Johnson of the University of California at Berkeley). Nor is the United States a reflection of greater reproductive success that came to our founding fathers as they drafted the Constitution. Rather, we are a nation of laws because our founders got together and, borrowing liberally from preceding examples, adopted a set of principles on how a society might and should be structured and run; the laws are simply formalized rules to ensure that those principles are applied. Our laws are blueprints for culturally defined canons of behavior. They are not blueprints of some hidden agenda, whether dictated by God or by genes.

Creationists do indeed believe that the very basis of our morals and ethics devolves from our having been created in God's image, and from our allegiance to that belief. Many ultra-Darwinians appear to believe that we must use our collective intellect to overcome the nastiness of the ruthless, competitive world portrayed in their version of Darwinian processes[1]—a radical form of "original sin" stretching back to the beginnings of life and the earliest bacteria over 3.5 billion years ago. I do agree that humans can and routinely do develop cultural norms that frequently supersede or override biologically based proclivities, but the ultra-Darwinian view that each one of us must fight the good fight anew against our basest, gene-driven instincts is probably not a very accurate picture of who we are or of how the evolutionary process has contributed to shaping the cooperative behavior that underlies all human social organization.

Lest you doubt that politics enters into private sexual and reproductive matters, think about the myriad laws covering abortion and the distribution of condoms and other forms of birth control. Early into the new millennium, the U.S. government has embarked on a quiet, but very real, campaign of disinformation about all manner of reproductive issues—revising official websites in an effort to appease the Christian right. The efficacy of condoms in preventing HIV, for example, has been questioned on a recently revised NIH website.

Conservative Christians tend to oppose evolution (though by no means unanimously), and no ultra-Darwinian, presumably, would be comfortable discouraging the use of condoms for either the prevention of pregnancy or the spread of venereal disease. But the doctrinaire overtones of hard-core, gene-centered evolutionary biology (and the insistence that the lessons of Darwinism be strenuously resisted in politics) are only a heartbeat away from some of the basic positions of the political right.

Biological determinism should not, I suppose, invariably lead to fascistic social polices and racism. But it often does. Anthropologists have been forced to abandon terms like "primitive" and "advanced" in their writings because "primitive" suggests "not as good as" "advanced" to many people—in the academic world as well as in society at large. In the rest of evolutionary biology, we use the words without fear in evolutionary analysis, with the precise meaning of "primitive" simply being "that which came before" "advanced." In paleontology, there is no "better" or "worse" attached to "primitive" or "advanced" (or its synonym "derived"). But anthropologists can't take that chance with such loaded terms.

For example, from where I sit as an invertebrate paleontologist, I have no problem seeing the hunter-gatherer way of life, the antecedent to settled farming existence, as the "primitive state" of humanity. I mean that until some 10,000 years ago *everyone* lived this way, still ensconced within the local ecosystem, along with all other forms of life that have ever existed on the planet over the past 3.5 billion years.

But I see the point: until the 1950s, the British government of the Bechuanaland Protectorate (today's Botswana) issued hunting licenses to exterminate San peoples—who, as hunter-gatherers, had no concept of private ownership of land or livestock, and so took what came their way as they had been doing on their open land since time immemorial. Black and white

Africans alike saw the San as subhuman, even making fun of their famous, very complex click language(s) as "monkey talk."

So a propensity—a cultural idea—seems deeply embedded in human life that other peoples are inferior. Obviously, the notion does not arise in science.

But when science comes along and talks convincingly about the importance of innate qualities as determinants of human behavior, there has been an inevitable temptation to characterize differences—whether between individuals within a family or within a town, or between ethnic groups living cheek by jowl (and thus competing for land and power), or peoples of different social strata, or peoples of different nations entirely—as themselves innate, now meaning "genetically determined."

I am not saying that all ultra-Darwinists, sociobiologists, or evolutionary psychologists are racists, fascists, or even political conservatives or social elitists. (Some of them, in resisting what they see as the potential implications of their version of Darwinism, are staunch liberals.) Nor am I saying that we'd better be wary of what we say—even if it is "scientific truth"—lest what we say fall in the wrong hands and unwittingly supplies the grounds for the social equivalent of an atomic bomb to malevolent people.

I *am* saying that we'd better be right about what we say in biology in general and about human beings in particular, for what we say has a long history of supporting social doctrine. The implications of supposed "scientific" demonstration of intelligence differences between "races" are incendiary:[2] so they'd better be correct.

There is a lot that is not correct about ultra-Darwinian evolutionary biology. Genes do not run the evolutionary process any more than they control all but the more elemental physiological aspects of my life. Social systems merge the economic and reproductive aspects of organismic life in fascinating ways; they are not just reproductive cooperatives. People do not hunt the way they hunt purely because success in the past in this mode of life raised the "fitness" of individuals who first started hunting that way millions of years ago; for culture was already on the rise, with its own rules, and providing learnable tricks and shortcuts to solve many problems in daily life—including putting dinner on.

Does this mean that evolutionary psychology falls like a house of cards, just because it is based on a flawed description of social systems, itself based on a distorted account of biological nature and the evolutionary process? No, not completely: there is some truth to all of these three disciplines—ultra-Darwinian evolutionary biology, sociobiology, and evolutionary psychology—if for no other reason than that humans have evolved and that some aspects of our behavior are under at least partial genetic control. And the gene-spreading premise of ultra-Darwinian evolutionary biology works well specifically for the very context to which its terms are best suited—systems and situations where out-and-out reproductive competition is taking place; the selfish gene works well enough for pure sexual selection.

But this trio of gene-based disciplines does not by any means give a complete and accurate account of who we are and how we came to be. If I'm right—that they are essentially cultural constructs devised, at least initially, to deal with the molecular revolution, far more than they are the simple fruits of honest, value-free toil in the lab, with desktop computers and the analytic minds of their scientific champions—then we had better be very careful in making assumptions that genes are as firmly ensconced in the driver's seat of life as these scientists, and many in the media who follow them slavishly, so ardently believe. The freedoms and quality of life of everyone are at stake.

## Summing Up

I have chosen to write this narrative in informal language, for though the issues at base belong to the inner sanctum of academe—especially to evolutionary biology—they affect us all: the extent to which we are creatures of our genes has enormous importance to everyone. Every day we are bombarded by a stream of reports, ads, and chitchat telling us how important genes are in determining who we are and what we do. We are assured that though the game of life may boil down to eating and mating (for us as well as every other living species), what really matters is the mating: we are out, in the last analysis, to spread our genes.

The arguments over the significance of genes and how the evolutionary process works spilled over the narrow boundaries of academe as soon as Darwin broached the idea (sooner, in fact). True, the issues are complex, but none of them are all that difficult to understand, and in using common language rather than the restrained and formal parlance of the academic world I hope to highlight just how easy it is to understand biological evolution and how it does and does not apply to human beings.

1. Evolutionary psychology, like the sociobiology from which it sprang, sees the need to spread genes as the organizing principle around which all of human society is constructed—and all human behavior is to be interpreted. Because life has evolved, and because the behaviors and anatomical structures of organisms that perform certain tasks are (for the most part) adaptations (bird wings, e.g.), ultra-Darwinian biologists have concluded

that all competition in nature—whether for resources or for mates—ultimately means nothing unless it affects the reproductive success of individuals. The modern notion of "fitness" in evolutionary biology was therefore defined purely as the probability of leaving offspring (thus genes) to the next generation, whether that reproductive success arises from success in the economic world (securing energy and nutrient sources—i.e., natural selection) or purely in competitive reproductive matters (i.e., sexual selection). Translated, this means that one may eat to live, but that one lives to reproduce, because evolution is so important.

2.  But life is not "about" evolving. Life is about living. Organisms are both economic "machines" and reproducers. Economics is essential to mere existence; reproduction is not necessary for the survival of individual organisms. Without reproduction, however, life would long ago have ceased: we reproduce not because we have to, simply to live, but because only the forms of life that did reproduce were the ones that managed to survive—not just the Pleistocene a million years ago but also the Precambrian 3.5 billion years ago.

3.  Economic and reproductive behavior automatically sets up separate, hierarchically structured systems of progressive inclusivity—systems devoted wholly to either economic behavior (the subject matter of ecology) or reproduction (the subject matter of evolutionary biology). The ecological hierarchy consists of organisms in populations, inside ecosystems, themselves interconnected to form larger, regional ecological systems and, eventually, the entire biosphere. Organisms are also parts of breeding communities (demes), which in turn

are parts of species—themselves the basal parts of the Linnaean hierarchy of life. Recognizing that these systems and corresponding disciplines exist in their own right is essential to a complete description of the entities and processes of the biological realm. It is essential, as well, to understanding how natural selection and other aspects of the evolutionary process work and what the nature of social systems might be.

4. Organisms do not live to reproduce, but if they are relatively successful at living, they may also reproduce; those best at making a living in general are more likely to reproduce. This is essentially the original Darwinian concept of natural selection; selfish-gene theory, in contrast, sees natural selection as an active process of genes (and the organisms that carry them) in constant competition to maximize the presence of those genes in the next succeeding generation. This is a rhetorical ploy to make genes seem like active players in the evolutionary game of life. They are not: genes are merely pieces of information, used largely to build bodies for economic purposes, whether self or offspring. Change in the genetic composition of populations (i.e., through natural selection) is, instead, a passive reflection of "what worked better than what" in the economic sector translated to the genetic ledger on the reproductive/evolutionary side of life.

5. But why sex? From the standpoint of gene spreading, sex is only half as efficient as asexual reproduction. Some biologists have argued that the presence of two sets of the same genes, one from each parent, allows repair of genetic damage and the possibility of alternate

gene expression: genetic variation in the service of the individual.

6. Perhaps even more critically, once established, sexual breeding systems—species—are very resilient, very difficult to get rid of through environmental perturbation, as they are broken up and integrated typically as local populations into a wide variety of ecosystems. Species resist extinction, often lasting for millions of years with little or no evolutionary change; asexual clones do not have such staying power, and that is probably the real reason why in most organisms sexual reproduction is far more common than asexual reproduction. Little evolution occurs without economic perturbation sufficiently great to get rid of a number of regional species. This demonstrates that evolution cannot be simply the result of the ineluctable urge for organisms to reproduce: local disturbance triggers ecological succession, but little visible evolutionary change. Mass extinctions, in contrast, reset the evolutionary clock and lead to the evolution of entire new groups— like modern corals—after their ecological equivalents have become extinct. Intermediate, regional turnovers (such as the one 2.5 million years ago in Africa, an event that affected human evolution along with many other lineages) involve the extinction and evolution of species, are common in the history of life, and are probably the locus of most evolution in the history of life.

7. Social systems are fusions of the economic and reproductive adaptations of organisms at the population level. Invertebrate colonies, including insect societies, are organized according to shared possession of genes, and division of labor within such colonies is very much like

cellular division of labor within the metazoan body. Even so, invertebrate social systems are economic and reproductive cooperatives—and not simply breeding colonies.

8. Vertebrates, as the overwhelming rule, do not share with insects special breeding systems with different degrees of genetic relatedness within local units beyond the existence of core family units. Selfish-gene theory, "inclusive fitness," and the like are thus concepts that do not apply well to vertebrate social systems.

9. Culture—learned, rather than genetically based instinctual, behavior—has long since dominated human ecology and reproductive behavior. As a result, human social systems are vastly more complex than any other known social system, because the economic sector of life, in particular, is organized into a very large number of hierarchical systems in which any individual (especially in postindustrial society) may be simultaneously a part.

10. The invention of agriculture beginning some ten thousand years ago effectively removed humans from the confines of local ecosystems—the context in which all the rest of life, including early phases of human evolutionary history, has always gone on. Natural selection takes place in local populations within local ecosystems, so the agricultural revolution (along with medicine and other cultural practices) has diminished the role and importance of natural selection in human biology. It also is the main reason why the human population has soared from five to six million to over six billion in these last ten thousand years.

11. Moreover, sex has become decoupled from reproduction

in human life, forming the "human triangle" of sex, reproduction, and economics. Though, as in all of life, there can be no reproduction without both sex and economics, sex exists for its own sake in human life and has, as well, intricate connections with the economic world quite apart from reproduction.

12. The human triangle of sex, reproduction, and economics—and the specific interactions between sex and economics and between economics and reproduction—while complex, can be teased apart and seen as repeated patterns. In particular, the human triangle works at the level of individuals (chapter 9) and also at the level of entire social groups (chapter 10). Sociobiologists and evolutionary psychologists often claim that they are inventing a "science" of human behavior—a more rigorous approach to the understanding of the origins, history, and current structure and function of the human condition than traditionally offered by the social sciences. Though often couched in conciliatory language (literally as in E. O. Wilson's "consilience")—where the supposed insights of gene-based evolutionary biologists are to be amalgamated with "other ways of knowing" of the social sciences—the program espoused by sociobiologists and evolutionary psychologists seeks in reality to reduce human existence (and its evolution) to the hard-core principles of ultra-Darwinian biology.

13. In contrast, social scientists themselves have long sought a more rigorous, "scientific" approach to understanding the regularities of human existence. They acknowledge that humans are animals and have evolved through processes that account for the evolution of all life in gen-

eral. But they maintain that culture—modes of behavior transmitted through learning, rather than genes (though the capacity for culture, everyone agrees, is genetic)—is often stronger than genetic influences. For example, sexual practices are known to be very malleable culturally; even genetic diseases can be fought culturally (through medicine and now also genetic engineering, which is, after all, a cultural manipulation of genetic systems).

14. Cultural determinants of human behavior frequently (though not invariably) run counter to patterns posited by selfish-gene theory—the main way that higher-level (i.e., cultural, though sometimes referred to as infrastructure) causality can be told from lower-level (internal, genetic) causes. The two (biological and cultural) processes may coincide (as they apparently do in some instances of female infanticide)—so that culture might even be interpreted as an enhancer of biology—but many examples show that the two are distinct and that, when they conflict, culture takes precedence over biology. China's one-child-per-couple policy is a simple case in point. More subtle are the repeated patterns of wealth and reproduction, where the general relation between economic and reproductive success (true for all of life since its inception, and the fundamental principle of natural selection) are usually reversed in human society.

15. Thus human life is the very worst candidate for blanket explanation in sociobiological/evolutionary psychological terms. There is little evidence for the assumption that much of human behavior evolved for purposes of hunter-gatherer life on the African plains millions of years ago, and that socially aberrant behavior such as rape reflects now dysfunctional holdovers from a primor-

dial time when rape was an efficient strategy for spreading (male) genes (the great Pleistocene cop-out).
Instead, most of the virtues and ills of contemporary
human life reflect cultural norms invented for increasingly crowded living in post–agricultural revolution
times of the past ten thousand years.

## FINAL THOUGHTS

In a book that appeared just before this one goes to press,
Richard Dawkins comments on what he sees as the main difference between his position and that of the late Stephen Jay
Gould.[3] He concludes that it is his own vision that genes play a
causal role in evolution, while Gould (and I, of course, as is
clear from this text) sees genes as passive recorders of what
worked better than what. Dawkins is exactly right—that *is* precisely the heart of the disagreement between ultra-Darwinians
and "naturalists" (for want of a better term) like Gould and
myself.

But I was astonished at the weak form of genic causality that
Dawkins throws up in castigating the genes-as-bookkeepers
metaphor as being exactly backward. True, he alludes to the
*causal* power of its own fate that a gene supposedly has, but that
seems to mean only that "whatever else genes are, they must be
more than book-keepers, otherwise natural selection cannot
work. If a genetic change has no causal influence on bodies, or
at least on *something* that natural selection can 'see,' natural
selection cannot favour it or disfavour it. No evolutionary
change will result."

How utterly disappointing! Dawkins seems to be saying that
people like Gould deny that mutations occur, and that varia-

tions of genetic information have different effects—different expressions—on the physiques, physiologies, and behaviors of organisms. How absurd! Ever since Darwin, all evolutionary biologists have accepted that natural selection is (as Ernst Mayr used to put it so forcefully) a two-step process: there is heritable variation (the ultimate source of which is mutation) (step 1), and what worked better than what in a finite world is what is kept (step 2). Gould believed that, I believe that—I don't know of a single evolutionary biologist who doesn't see it that way. And that, as we have seen, was precisely the way Darwin saw natural selection, despite his ignorance of genes.

But rather than the fleeting, alarmingly phantasmagorical thought that I have been tilting at windmills here, I was relieved to see, in an essay in the same book, that Dawkins at least originally was up to a much stronger sense of how genes control their fate. The telling words are not actually his, but those of George Williams, albeit clearly quoted with approbation: "With what other than condemnation is a person with any moral sense supposed to respond to a system in which the ultimate purpose in life is to be better than your neighbor at getting genes into future generations. . . ."[4]

I am relieved. Whether it is the individual (human, or anything else) or the genes themselves (Dawkins's original metaphor from the 1970s), the causal activity attributed to genes far transcends the bland and uncontroversial presentation of heritable phenotypic variation to the relentless maws of natural selection.

No indeed—the gene-centered version of the universe, created in part and deftly articulated by Dawkins, is out for far more causal blood than the mere sending out of variants to the generational selectional firing line. Those of us who have looked in vain for the strong imprint of the selfish gene on the history of life; those of us who have, above all else, been persuaded that nothing much of interest happens in evolution unless provoked by physical environmental stimuli; those of us who see the ultimate causal vector in evolution to be the environment, and *not* the insatiable desire of anything, organisms or their genes, to spread their genes/themselves—we can all rest easy in our belief that genes, the blueprints for making all organisms, are indeed the cumulative record keepers of what has worked better than what, as determined by natural selection (and some random hits). That's all it is, though that's a lot. But it is only that—and nothing more.

# Notes

*Chapter 1: Obsessed with Genes*

1. The recent sex abuse scandal in the U.S. Catholic Church to the contrary notwithstanding.

2. By Michael Pollan, July 19, 2002, p. A17.

3. Though I am a paleontologist (and trained originally in a geology department), I was nonetheless the prime mover in bringing the first molecular biology laboratory into the American Museum of Natural History, in the 1980s. I identified space—and was willing to use a vacant research position and argue for the appropriate budget—because it was clear that the general enterprise of analyzing evolutionary relationships ("systematics"—the prime focus of biological research in all natural history museums) was entering an exciting new phase utilizing genetic data. So I am no knee-jerk, anti-gene paleontologist, but rather an evolutionary biologist seeking a more reasoned balance when it comes to evaluating the factors that contribute to the evolutionary process.

4. And Nobel laureate—let us never forget that science, like many another form of human endeavor, is highly competitive. Stakes are high, however, since in academe generally, it is power and glory, rather than conventional personal wealth (though size of labs and degree of internal and external funding of the enterprise are critical), that are at stake. A rich Kyoto Prize is sometimes awarded to an evolutionary biologist (E. O. Wilson, for example, was once a recipient), but biology, let alone evolutionary biology, just wasn't on Alfred Nobel's original dance card when he established the Nobel Prizes,

which remain far and away the most prestigious (and monetarily significant) awards for which scientists qualify. Physics and chemistry (actually, a distant second to physics) remain the "real," "hard" sciences, while biology and geology tag along farther behind, followed by the even "softer" fields of anthropology, sociology, and the like. (Interesting, in this respect, that economics was on Nobel's original list!) Molecular biology qualifies for Nobel Prize treatment because it studies the chemistry of the molecules of inheritance (easily the hottest topic in science of the mid-twentieth century), and could be shoehorned into the Nobel category of "physiology and medicine." The near-universal reverence shown to recipients of the Nobel Prize (jealousy apart) is by no means confined to the public at large (the vast majority of whom don't really know or care much about such things); the reverence is actually most acute within science itself. And thus when Nobelists like Monod began to consider evolution (and in an intelligent manner, unlike so many other Nobelists who have dabbled in evolutionary biology—William Shockley, for example), evolutionary biologists themselves simply *had* to respond.

5. The term "molecular drive" was coined by the Cambridge molecular biologist Gabriel Dover in 1982, to describe how parts of a gene lying on a chromosome can change the form of its equivalent on the other, paired ("homologous") chromosome. As a potential source of bias in the transmission of genetic information, molecular drive qualifies as a legitimate potential evolutionary mechanism.

## Chapter 2: Chickens and Eggs

1. As when Onan shed his *seed* upon the ground.

2. Hence all the excitement about Dolly, the cloned sheep, whose creation marked the first time in either nature or the lab that the genetic information from a body cell (an udder cell, to be exact) was ever the basis for forming a complete biological individual—and the exception that proves Weismann's insights.

3. As had Alfred Russel Wallace, who had independently formulated the notion of natural selection before Darwin published his ideas. The two presented a joint paper on natural selection to the Linnaean Society in London in 1858. It was Darwin's treatment of natural selection in his *On the Origin of Species by Means of Natural*

*Selection* (1859), however, that really established both the nature and the validity of natural selection as the cornerstone of the evolutionary process.

4. The modern notion of "fitness" seems to have arisen in the early days of population genetics, which is a predominantly mathematical approach to the study of the factors that alter gene frequencies in populations and species. Fitness is usually rendered as $w$, which is equal to $1 - s$, $s$ being the "selection coefficient." Thus it seems entirely possible that the muddying of the distinction between economics and reproduction in the original Darwinian formulation of natural selection originated as a computational convenience rather than as an improved description and understanding of the process itself.

5. Bacteria reproduce by dividing, and thus an individual bacterium is potentially immortal. But given the hazards of life, no individual, not even a bacterium, is likely to survive very long—let alone forever.

6. The metaphor of the selfish gene has a striking parallel in the even more phantasmagorical scenarios of the development of artificial intelligence. I once heard a geneticist in an academic setting seriously propose the threat that robots may eventually take life into their own hands—making more of themselves according to blueprints (i.e., not genes, but rather the way people today construct material cultural items) and take over the world! I sat there wondering why people couldn't just pull the plug, simply disrupt whatever energy source the robots were using. But at least artificially intelligent robots might be construed as having a conscious sense of purpose—of *wanting* to make more of themselves. How bits and pieces of genetically coded information can be construed somehow to be in a race to leave more copies of themselves to the next generation seems, in this respect, even harder to understand.

7. Perhaps not so uniquely after all, as once again the parallel with religious doctrine (especially the church) comes to mind— where the do's and don'ts, the oughts and ought not's, in the relation between sexual behavior and reproduction, have been the focus of unending analysis and disquisition.

8. I.e., strictly in terms of genes; time and energy are always "costs" in reproduction.

## Chapter 3: *The Natural Economy*

1. The energy powering all ecosystems comes from the sun—the sole exception being the vent faunas of the deep sea; their energy derives from heat flow from radioactive decay deep within the earth.

2. For example, Richard Dawkins writes in *The Selfish Gene* (Oxford: Oxford University Press, 1976), p. 90, "Maynard Smith's concept of the ESS ['evolutionary stable strategy'] will enable us, for the first time, to see clearly how a collection of independent selfish entities can come to resemble a single, organized whole. I think this will be true not only of social organization within species, but also of 'ecosystems' and 'communities' consisting of many species. In the long run, I expect the ESS concept to revolutionize the science of ecology."

## Chapter 4: *The Consequences of Baby Making*

1. Alcohol is the most notorious and, some would say, salubrious by-product of yeast metabolism. That alcohol is a poisonous substance that yeast cells must get rid of to survive is nicely borne out by the typical 13–14 percent range of alcoholic content in a bottle of wine: fermentation ceases when alcohol reaches this level, because alcohol at this concentration is lethal to yeast cultures. On a related point, consider how a mammal body differentiates—making different cell types, tissues, organs, and organ systems—and grows: also by cell division. The "individual" animal is actually a gigantic colony of genetically identical interacting cells. So we're not all that different from yeast after all (though the alcohol we make comes from tamed colonies of yeast in vats, not from our own cells).

2. Each coral polyp in a colonial coral system is genetically identical to all the others—much the way individual yeast cells in a culture, and all the cells in a mammal's body, are genetically identical. This leaves mutation to be the sole cause of genetic differences between cells in a culture, in a colony, or in an animal's body.

3. The South African geneticist Hugh Paterson coined the term "specific mate recognition system" (SMRS for short). That, in effect, is what a species is: a collection of dynamically interbreeding individuals sharing a mate recognition system.

4. In the final vignette of his film *Everything You Always Wanted to Know about Sex . . . But Were Afraid to Ask*, Woody Allen plays a sperm sitting around with other sperms (sort of like the way parachutists line up before the jump) about to be ejaculated (if, that is, the brain can sustain the erection—which is touch and go on this particular date). The sperms basically don't want to go, and they commiserate with one another. There is little in the way of we-gotta-get-out-and-get-there-first competition in Allen's depiction of the mind of the sperm!

5. Mistakes occasionally happen. When visiting a research facility once, I learned the rather poignant story of a young lazuli bunting female. She was born in the Midwest in one of the few places where lazuli and indigo buntings breed side by side. Though a lazuli bunting in all respects—she was no hybrid—she had learned the song of the male singing in the next-nearest nest, who happened to be an indigo bunting, and not her lazuli bunting father. She was destined never to raise lazuli chicks of her own. Birds can recognize the "song of their species," but also tell the difference between individual males. And sometimes the males just never get the song right. It was disturbing, for example, to listen to a white-throated sparrow singing twenty-four hours a day one recent spring. Males normally do not sing at night, and this poor guy never sang the otherwise fairly stereotypical song that white throats sing to define their territories and find their mates.

6. See *The Sibley Guide to Birds* (New York: Alfred A. Knopf, 2000) for details of this chickadee story, as well as for background information on all other North American bird examples in this and other chapters.

7. The great geneticist Sewall Wright examined the semi-independent histories of localized populations (he coined the word "deme"), especially in his "shifting balance theory" of the early 1930s. A major problem in understanding evolution, as Wright saw it, was how favorable genes could spread from deme to deme—and thus the genome of a species would change through time. He spoke of "genetic drift"—where versions of genes ("alleles") could become "fixed" in a population simply through the vagaries of the genetics of the reproductive process, and not necessarily through the action of natural selection. Later, the geneticist Motoo Kimura and others developed the "neutral theory," which is based on the observation that many alleles are "selectively neutral"—i.e., are neither beneficial nor harm-

ful to an organism when compared with other forms of the same gene—offering further insight into how genetic diversity is established both within and between populations.

8. Most paleontologists have long since come to agree that stasis—the strong stability of species without accruing much, if any, noticeable evolutionary change, over most of their existence (which in invertebrates can be five or ten million years)—is a very real, and significant, feature of the evolutionary history of life. The cornerstone empirical observation underlying the idea of "punctuated equilibria," stasis has in the last decade or so begun to be accepted by evolutionary geneticists as a problem that needs to be explained—the arena now changing to competing notions of what causes stasis. Nonetheless, there are still holdouts who deny the phenomenon of stasis, even within the ranks of paleontology, and certainly among geneticists and other evolutionary biologists who are obsessed with the gene-centered picture of the biological universe.

9. As we have seen, Sewall Wright coined the term "deme" for local breeding populations. Years later, the paleontologist John Damuth used the word "avatar" for local populations of organisms of the same species that are parts of (have niches in) local ecosystems. Avatars and demes may be similar in composition, but are generally not the same.

## Chapter 5: Economics + Babies + Time = Evolution

1. Readers might assume that I am leading into a discussion of the (rather arcane) topics of "group selection" and "species selection" when I say that the sort of interactions between the genealogical and the economic hierarchies seen in the dynamic feedback loop of baby making/natural selection within local populations is "echoed" at higher levels. Though group and species selection (not, by any means, the same issues) are legitimate topics in evolutionary theory, they are side issues for the purposes of this narrative, which simply seeks to spell out the consequences that come from the separable aspects of economics and reproduction in any organism's life—including (eventually!) why people have sex.

2. Many ecologists have argued that fire is not only natural but in a sense necessary to the long-term health of terrestrial ecosystems—not least because, in this triggering of ecological succession,

soils are replenished and (often) the diversity of different species, hence the complexity of the ecosystem, is increased, thus leading (again, in the view of many, if not all, ecologists) to the continued stability of the system as a whole. Against this view of the beneficence of fires is the short-term degradation of nearby human life; humans almost universally view burned-out landscapes as blights (makes sense—ancestrally, forests, and grasslands were our homes); burned-over slopes are prone to erosion and even landslides; and loss of homes is downright devastating. This sets up an interesting dilemma: a good way to prevent the spread of wildfires is to clear the highly combustible underbrush and dead trees from a tract; but environmentalists are understandably concerned when George W. Bush enthusiastically endorses "selective logging" in National Forest and other U.S. government-owned lands, ostensibly to reduce fire hazard, but not coincidentally allowing logging interests easy, cheap access to the rich wood resources on public lands.

3. It's been said that, without certain species of termites, with particular kinds of cellulose-attacking microbes ensconced in the hind gut, or alternatively with fungal farms inside their mounds, there would be little or no decomposition of plant materials in some of the drier environments of the tropics.

4. Ferns come back notoriously fast after forest fires, because their spores, already in the ground, are quite resistant to heat. That's one of the reasons geologists determined that massive fires raged around much of the globe after the cometary impact that killed off the dinosaurs and many other groups at the end of the Cretaceous period, 65 million years ago: the soils in surviving terrestrial deposits are extra rich in fern spores (the so-called fern spike) just after the event occurred.

5. This is a gentle, subtle example of the general phenomenon of "historical contingency," a major theme in the evolutionary thinking of the late Stephen Jay Gould. Gould had his mind mostly on larger-scale phenomena, like mass extinctions, which we'll get to shortly. But the main point is that, no matter how similar the recurring patterns might be, their details always differ—and the greater the magnitude of the shock wave (i.e., environmental event) that triggered the evolutionary response, the more the differences are magnified.

6. Darwin discusses species as "permanent varieties" in *The Descent of Man*, the same work in which he gave us the clear distinction between natural and sexual selection.

7. See my *The Pattern of Evolution* (New York: W. H. Freeman, 1999) for an account of the resonance between the development of an evolutionary theory of Earth in geology, and biological evolutionary theory. I still believe that Darwin (whose most intense formal training in science was actually in geology) did more than anyone else to establish the correct order of magnitude of time for the age of Earth—this in an era long before the discovery of radioactivity (and thus radiometric dating). Darwin needed the time for his notions of biological evolution to produce the long history, and great present diversity, of life on Earth.

8. And North America. Agassiz later became the founding director of Harvard University's Museum of Comparative Zoology. Though he opposed Darwin's ideas on evolution, Agassiz was a great zoologist and geologist, carrying out much important descriptive and interpretive work after he came to America.

9. Not to overstate the case: most of the famous Ice Age mammals appear to have evolved on the steppes of Siberia around the onset of the second glacial, which happened roughly 900,000 years ago. It was humans, however, who appear to have been directly responsible for their demise—at least in North America, where there is a striking correlation between the appearance of humans (at least in any significant numbers) some 12,500 years ago, and the soon thereafter disappearance of mammoths, mastodons, and the rest of the Pleistocene megafauna.

10. The newly revised date for the base of the Cambrian is 543 million years ago, down from the 570 million years I learned in school, and a date that gives considerably less time for the initial stages of life to have been played out than had been previously thought. The jury is still out deciding whether the animal phyla—the arthropods, mollusks, annelid worms, echinoderms, etc.—actually first evolved at the base of the Cambrian, where first we find them. Alternatively, some paleontologists prefer to think that the animal phyla diverged perhaps as far back as 1.2–1.3 billion years ago, on the basis of estimates derived from a comparison of the genetic sequences among these major groups. Either way, the striking pattern of relatively abrupt appearance and proliferation of trilobites and other "shelly" marine invertebrates, more or less at the same time as what geologists accordingly defined as the base of the Cambrian system (lowest major division of the Paleozoic era) is a sure sign of a major evolutionary spurt. If the forebears of these phyla had indeed been around for any length of time prior to

that, they must have been tiny and without the hard mineralized skeletons that make organisms ripe candidates to end up as fossils. The Cambrian explosion thus by definition entailed a tremendous amount of anatomical evolutionary change—probably triggered by major environmental events (such as the famous "snowball earth," when the world was repeatedly glaciated way down into the tropics several times just prior to the "explosion"). It is also likely that there was insufficient oxygen (released by photosynthesis) in the world's oceans to support macroscopic multicellular animal life until a threshold was reached near the base of the Cambrian—a conclusion based on geochemical evidence decades ago (the "Berkner-Marshall hypothesis") that has held up well over the years.

11. I am relying in part on my personal experience with these fossils, but especially on the publications of the paleontologists Carlton Brett, Gordon Baird, and others for these statistics.

12. The American Museum of Natural History and Columbia University paleontologist Norman D. Newell called mass extinctions "crises in the history of life." Mentor to myself, Stephen Jay Gould, and many other students, Newell stood virtually alone among paleontologists in the mid-twentieth century in pointing to the reality—and huge evolutionary significance—of global mass extinctions. It took all of us decades to catch up with him.

13. This is not to suggest that *Triceratops*, that tank-like horned herbivorous dinosaur that shared the last gasps of Cretaceous air with *Tyrannosaurus*, was replaced, say, by rhinoceroses. The point, rather, is that four-legged vertebrates dominating the world's terrestrial ecosystems were predominantly dinosaurs and other reptiles in the Mesozoic, and predominantly mammals in the Cenozoic. How similar the plants and animals will seem to be before and after an extinction event depends on the nature of the genetic information surviving the event. When Paleozoic corals succumbed 245 million years ago, corals evolved again—this time from the sea anemone group, some of whose members evolved the capacity to secrete calcium carbonate to form a rigid skeleton. Absent true corals, other calcareous organisms (some algae, sponges) can also form reefs, but Mesozoic and Cenozoic reef systems look much like their Paleozoic counterparts because the corals in each were, in fact, quite closely related to one another.

## Chapter 6: Clones, Colonies, and Social Life

1. The milestone statement that the evolutionary process essentially boils down to competition for reproductive success—and spread of genes—came in George Williams's highly influential *Adaptation and Natural Selection* (1966). A decade later, Richard Dawkins's *The Selfish Gene* (1976) took the argument a step farther down—with his metaphor of genes, rather than organisms, competing for survival to the next generation. But it was the Oxonian Bill Hamilton (recently deceased from malaria contracted in the field) who wrote the seminal papers, published in 1964, that established the correlation between degree of cooperation and genetic relatedness among organisms. Rumor has it that one of the founders of mathematical population genetics, Ronald Fisher, had come to much the same conclusion some thirty-odd years earlier—according to legend performing quick and dirty calculations on an envelope in a British pub.

2. The Michigan State biologist Richard Lenski, with students and colleagues, has kept colonies of *Escherichia coli* going for over thirty thousand generations, providing perhaps the richest database on evolutionary phenomena ever to come from a laboratory.

3. The soma, the economic side of any animal's body, is also a hierarchically arranged system: cells are parts of tissues, which are parts of organs, which are parts of organ systems—a simple extension down and inward of the economic hierarchy we encountered earlier—which starts with organisms and works its way up through avatars, local ecosystems, and so forth.

4. As we shall soon see, when it comes to the vastly greater complexity that human social systems represent, there are distinct advantages in some human settings to seeing one's own genes make it to the next generation—yet these advantages are purely economic!

5. See E. O. Wilson, "The Sociogenesis of Insect Colonies," *Science* 228 (1985): 1489–95.

6. That hives may be closely packed together, and their component bees rubbing shoulders with one another while out foraging, just means there is a hierarchical fine structure to the way populations of a given bee species act locally within an ecosystem.

7. Interestingly, however, at least some experts on lek behavior

are concluding that more is going on than just competition for mates on the lek dancing grounds. For example, in a *New York Times* article ("Is Dancing Chicken Looking for Mate, or Looking for Cover?," April 29, 2003), the biologist Robert Gibson of the University of Nebraska is quoted as saying that, in addition to looking for mates, male greater prairie chickens are "trying to stay alive . . . by joining a group to evade predators."

8. As in the old story about the old bull and the young bull, standing on a hill overlooking a herd of cows contentedly grazing the grassy sward beneath their feet. Young bull: "Whaddya say? Let's run down there and jump a coupla them cows!" Old bull: "Son, let's walk down and jump 'em all!"

9. G. E. Woolfenden and J. W. Fitzpatrick, *The Florida Scrub Jay: Demography of a Cooperative Feeding Bird* (Princeton: Princeton Univ. Press, 1984), a study that remains the authoritative word on the subject of Florida scrub jay social life.

10. E.g., R. W. Wrangham, "Ecology and Social Relationships in Two Species of Chimpanzees," in D. I. Rubenstein and R. W. Wrangham, eds., *Ecological Aspects of Social Evolution: Birds and Mammals* (Princeton: Princeton Univ. Press, 1986), pp. 352–78, which I follow here, and Frans de Waal, some of whose insights on issues in evolutionary psychology appear later in this narrative.

11. The great biologist Lynn Margulis is best known for this line of thought, beginning with her demonstrating to everyone's satisfaction that the eukaryotic cell arose as a fusion of two separate (perhaps more) kinds of bacterial cell: the mitochondria and plastids (the metabolic power plants found in animals and plants, respectively) have a separate single-stranded complement of DNA—DNA that has nothing to do with the DNA housed, in double complement, in the nucleus of the cell.

## Chapter 7: Naught So Queere as Folke

1. Recent discoveries make it clear that earlier African hominid species had already established a beachhead in Eurasia; yet the evidence (in the form of stone tools) also makes it clear that hominids were not present in great numbers until the onset of the second major glacial advance of the Pleistocene "Ice Age," just under a million years ago.

2. Hunter-gatherer peoples (e.g., the San, or "Bushmen," and the BaAka and Ba Mbuti ("pygmies") of Africa, and other groups in South America and Asia, who never adopted agriculture, have persisted into the modern era—though none survive culturally intact any longer.

3. Sickle-cell anemia is a medical problem still plaguing many people in Africa, the Mediterranean region, and the United States. It is prevalent where malaria is rife: the gene for sickle-cell anemia (a fatal disease) in its pure ("homozygous") form confers the disease; individuals with the alternate, non-disease-forming form of the gene ("allele") don't get the disease, but they are far more susceptible to catching, and dying from, malaria than individuals with one allele each—people who are resistant to malaria and do not develop the genetic disease sickle-cell anemia. Thus natural selection cannot eliminate the sickle-cell gene, because in its diluted form it conveys resistance to another killing disease.

4. According to the paleoanthropologist Ian Tattersall, in *The Last Neanderthal* (New York: Macmillan, 1995), p. 158, a Neanderthal individual with a withered arm, found at the Shanidar cave site of northern Iraq, "suggests that such groups afforded long-term support for at least some disadvantaged members."

5. Some groups of doctors, I hear, while attending seminars on longevity, are shown slides of the outlines of human evolution—from apelike ancestors to people up through fifty years into the future, the last one being "modern man"—portrayed not as a thin, emaciated E.T. but as an obese figure with relatively smaller heads; heads remain the same size, of course, and only look smaller in relation to the girth.

6. Obesity is quintessentially an environmental problem—caused by too much ready availability of high-fat, high-carb foods coupled with couch potato lifestyles. But this is not to say there is no genetic variation within the human population pertaining to obesity, because some of us can stoke up on Big Macs and beer without gaining an ounce, while others of us begin to look like the last slide in the doctor's presentation of the preceding note.

## Chapter 8: Sex Decoupled

1. Or so says Barbara Ehrenreich in her essay "The Oral Sex Solution," *Funny Times*, November 2000, p. 5.

2. The psychologists Steven Scher and Frederick Rauscher, in their introduction to their edited volume *Evolutionary Psychology: Alternative Approaches* (The Netherlands: Kluwer Academic Publishers, 2002), have drawn a useful distinction between "narrow" (I would say "hard-core") evolutionary psychology, which hinges on selfish-gene adaptationist evolutionary biology, and more general approaches to the evolution of human behavior. These approaches include well-established principles within psychology itself, as well as alternative formulations of the evolutionary process (such as, but not restricted to, those developed in part 1 of this book), and, I think critically, interpretations of the nature, structure, and evolution of social and cultural behavior by anthropologists and other social scientists.

3. Incredibly, critics of one or another claim of hard-core evolutionary psychology often find it necessary to swear that *of course* they believe that human behavior has evolved—and therefore has, to one degree or another, a genetic base. It is as if psychologists and anthropologists found themselves having to defend their status as scientists—i.e., as objective observers and analysts of natural phenomena, preferably with the aid of data (preferably experimental, or otherwise objectively captured)—just because they have what seem to them perfectly plausible, non-adaptationist descriptions and analyses of the same phenomenon, drawn from the canons of research and analytic theory of their own disciplines, instead of those of the interloping hard-core evolutionary psychology. This is none other than Wilson's "consilience" gambit, which we encountered in chapter 1—where ostensibly knowledge from these fields is to be unified, but in reality is to be consumed by the overarching theoretical framework of hard-core ultra-Darwinian principles. For an accessible account of the "jealousy wars," see the article by Erica Goode in the *New York Times* Science Times, October 8, 2000, p. F1, replete with defensive protestations by the critics of this evolutionary psychology scenario of jealousy, to the effect that of course they are not anti-evolution per se, and the self-inflated retorts by some of the evolutionary psychologists involved—who accuse their critics of "having agendas" arising from the "mudded masses of mainstream psychology," dismissing all criticism in quoted statements like "People have always been resistant to evolution. . . . We're in the midst of a scientific revolution in psychology." One can only assume such grandiose hyperbole is calculated to increase this guy's fitness. But the example also shows that

humans can be and routinely *are* jealous over other matters arising from the social contract—like who gets the best press, and commands the most money and campus respect—a point hinted at by the writer of the *Times* article.

4. A joke, of course—but in the old days of experimental, breeding genetics, researchers were very concerned with the degree of "penetration" of a gene, meaning how strongly it would be expressed in the phenotype of the organism.

5. See Kenneth Bock's *Human Nature and History: A Response to Sociobiology* (New York: Columbia University Press, 1980).

6. It is extremely important to note here that I am not about to advocate some form of "group selection" as an explanation of human behavior. Far from it! I am merely using the solid analytic approach advocated by George Williams and others to tell the difference between group-level and individual-level selection—a technique that I believe can be used with some success to tell the difference between genetically versus culturally determined traits and characteristics of human populations. This disclaimer is all the more necessary because I was an early framer of the notion of "species selection"—an idea that saw the differential survival of species as important to generating certain kinds of evolutionary patterns (evolutionary trends, specifically) commonly encountered in the fossil record. Species selection is different from what biologists meant by group selection: group selection, for example, held that selection between groups is based on "group mean fitness"—a concept that has no counterpart on the literature on species selection. In any case, I reiterate: *I am not advocating group selection, species selection, or any other form of selection as an alternative or additive to the concept of natural selection.* I am merely looking for a way to tell genetically based, evolutionarily derived behavior patterns from culturally imbued behaviors and practices.

7. Another example: though the calculations and physical evidence are there to support the notion of global warming, Earth regularly warms up to a far greater degree than the global mean temperature of Earth has been these last fifty years—and does so through "natural" means. Thus, while the evidence is firmly on the side of human-induced global warming, it may be working in conjunction with other, "natural" warming factors, and that makes the job of teasing the two apart all the more difficult.

8. In brief, Lewontin showed that there was a mechanism distorting sperm production in populations of house mice—so that a dispro-

portionate number of sperms carried this "*t*" locus, even though in the homozygous state it was either lethal or produced sterile males. "Deterministic models" of natural selection indicated that the stable frequencies the alleles should be found in natural populations, but Lewontin consistently found the alleles at lower frequencies in wild populations. He produced a selection model consistent with "group selection" that predicted more accurately the frequencies actually found in nature. See Williams, *Adaptation and Natural Selection* (1966), p. 117 ff.

9. See, for example, the anthropologist Marvin Harris on these points, especially his *Cannibals and Kings: The Origins of Cultures* (New York: Random House, 1977) and *Cultural Materialism: The Struggle for a Science of Culture* (New York: Random House, 1979). Harris, like the biologist George Williams in a completely different sense, finds the evidence that individuals can be readily socialized into accepting radically different norms of sexual, reproductive, and other behavior so compelling that it must be true that culture in general overrides biology as the causal agent of human behavior. Williams, recall, used a similar argument—the dearth of counterexamples plausibly attributed to group selection—to claim the ubiquity and primacy of individual-based selection. That culture so readily overrides so much of the alleged primal, all-important drive for humans to make babies is the main reason that some evolutionary psychologists balk at insisting that all human behavior we see today is necessarily still "adaptive"—that everything from buying cars, planting potatoes, and going to church is all about increasing individual "fitness" values.

10. Pure kinship systems (extended families, clans, etc.) are expressly genealogical, while bands, villages, towns, and larger geopolitical entities are not. Though xenophobia is, if anything, increasingly common in a world of expanding population and shrinking per capita resources, there are, of course, no "pure" ethnic societies. Jews undoubtedly share a genealogical as well as cultural heritage, but Israel is far from genealogically pure (efforts to deny citizenship to Ethiopian Jews to the contrary notwithstanding). Finland, which until recently prided itself on its ethnic monotony, has seen fit to be more open to immigration by peoples of more diverse ethnic backgrounds. Interestingly, people like the San of the Kalahari—until recently, at least, among the last of Earth's hunter-gatherer peoples— still see their bands as parts of local ecosystems, and thus fail to rec-

ognize either international boundaries (borders between Namibia, Botswana, and South Africa) or private ownership of livestock. Many Bushmen have been unceremoniously murdered by black and white farmers alike for not seeing the invisible-to-Bushman-eyes political boundaries and rules of ownership.

11. Daily migration for work goes back several centuries in Europe—and is mostly associated with cities (farmers still commonly live where they work, though corporate farming is changing that as well). And though workers can increasingly "mail it in" via the Internet—thus stay at home to work—it will remain largely true that the salary-providing businesses/corporations will have a locus (frequently many) where they are headquartered and administered and where goods and services will continue to be produced.

12. Lawn mowing on a regular basis is now codified in town ordinances in many New Jersey communities.

13. I admit the mind boggles a bit connecting lawn mowing with a romp in the hay, but given the endless creativity of the human mind when it comes to sex, I'll bet grass cutting has led to sex more than once—even in suburban New Jersey!

## Chapter 9: Up Close and Personal

1. See Clifford Geertz's review and analysis of the English translation of a book on the ethnography of the Na written by the French-trained Han anthropologist Cai Hua, in *New York Review of Books*, October 18, 2001, pp. 27–30.

2. Dr. Spock to the contrary notwithstanding, babies don't come with instructions. A case can thus be made that child rearing by specialized "experts"—or at least by people with some prior experience and a track record of success—is perhaps not a bad idea. The downside, of course, is the absence of the elemental, primal bonding and the type of emotional commitment—love—that only parents (and especially mothers) can give—that is, when they are up to the job.

3. According especially to the bonobo expert Frans de Waal.

4. Great apes, too, have long gestation periods (orangutans, for example, also require a full nine months)—meaning that taking the better part of a year from conception to birth is not newly evolved in the human lineage.

5. I first encountered the food-for-sex scenario in the anthropol-

ogist Helen Fisher's book *The Sex Contract* (1982). Though she herself has subsequently adopted more explicitly sociobiological views, I find her insights in *The Sex Contract* still compelling.

6. As far as I can recall, Aristophanes' theory does not explain bisexuality—which is actually a good thing, because theories that explain everything are usually suspect.

7. I do know of at least one older male who had a number of affairs and who produced children not only in his original nuclear marriage/family but also with at least two of the other women he had become involved with. Though he is the only male I know of who endorses the general philosophy that it is good to leave as many copies as possible of your genes to the next generation, he also agrees that sex is not *just* about making more and more babies.

8. The one case I read about actually involved a player on a team who was known for his voracious sexual appetite. The idea there was that, since so much of the guy's self-esteem was tied up in having lots of sex, it was best simply to keep him happy as the game approached!

9. Nor is this phenomenon associated merely with athletics. Men can feel invigorated by sex, but also depleted. That is probably why General Jack Ripper (Sterling Hayden) tells Group Commander Mandrake (Peter Sellers) that he likes women, but adds, "I do deny them my essence"; Ripper had just sent off a fleet of B-52s to drop nuclear bombs on Russia, in Stanley Kubrick's film *Dr. Strangelove*.

10. One of my all-time favorite cartoons, seen when I was but a callow, baseball-mad youth, showed a player saying into the microphone thrust into his face, "My biggest thrill??? I guess it was last year in Cleveland. I pitched, and we won, and after the game these two girls invited me up to their apartment. . . ."

11. "Do you want that vagina in chocolate or marzipan?" asked the clerk of his client on the phone in one of these emporia as I was stepping inside to check the place out—Upper West Side Manhattan, in the early 1980s.

## *Chapter 10: Sex, Economics, and Babies in Social Systems*

1. Some birds do use their wings to stabilize themselves as they copulate. Ostriches, with their reduced, rudimentary wings, don't—and have found other means to achieve copulation.

2. There are indeed diehards who insist that fidelity in vacuum

tube radios is superior to that in transistor radios. But transistors and microchips are smaller and less costly—and have made possible the computer revolution—clearly a case of "selection" in action.

3. Most notorious is the book *A Natural History of Rape: Biological Bases of Sexual Coercion*, by Randy Thornhill and Craig T. Palmer (Cambridge: MIT Press, 2000). See the review of this book by the primate behaviorist Frans de Waal, in the *New York Times Book Review*, April 2, 2000, for an especially sentient take on the book's thesis that "plain old evolution explains why men rape." The editorial heading ("Survival of the Rapist") encapsulates the gist of the book's argument. I couldn't agree more with what de Waal has to say, and my agreement is reflected in my discussion.

4. As de Waal points out (see reference in note 3), a penis is not a fist.

5. The Human Relation Area Files are archives of ethnographical research centrally stored at Yale University. Recently, some (but by no means all) of these files have been digitized and put on line, and it is to these digitized files that I refer.

6. This is true in Northern Ireland, Israel, and anywhere where identities are based on such nonphenotypic properties as religion. The generic cartoon about this shows a soldier who challenges an approaching couple strolling down the street and demands to know whether they are "friend or foe."

7. Kottak's book is *Assault on Paradise* (New York: Random House, 1983). The town is Arembepe, some forty kilometers north of São Salvador da Bahia. I was there as an undergraduate trainee in anthropology in the (northern) summer of 1963. I shared a rental house with another student—and "Dora" (not her real name) was our housekeeper—so I find Kottak's account (based in part on the paper his wife, Betty, published on Dora's life) a compellingly accurate portrait of the woman and her children.

8. Arembepe is where it is because a sandstone formation (loaded with invertebrate fossils, and a haven for living mollusks and sea urchins that were harvested at low tide) cropped out at the ocean, creating a protected harbor with a convenient breach for access to the South Atlantic.

9. Headline on the front page of the *New York Times*, November 27, 2002: "Women Catch Up to Men in Global H.I.V. Cases."

10. According to the United Nations Food and Agricultural Organization, as reported in the article cited in note 9.

11. Statistics drawn from *Pocket World in Figures, 2002,* published by the *Economist.*

12. Here and in later discussions of infanticide as a socially condoned, but seldom explicitly acknowledged, means of ex post facto birth control, I am relying on Marvin Harris's data and analysis in *Cannibals and Kings: The Origins of Cultures* (New York: Random House, 1977) and on Sarah Blaffer Hrdy's *Mother Nature: A History of Mothers, Infants, and Natural Selection* (New York: Pantheon Books, 1999).

13. According to Celia Dugger, writing in the *New York Times,* May 6, 2001.

14. See Hrdy's discussion in *Mother Nature.* The recent revelation that some cases of SIDS are infanticide—especially serial cases involving several infants born to the same mother—of course adds an extra burden on families that have lost infants to SIDS that clearly did *not* kill their children through neglect, whether deliberate or not. No one is saying that SIDS is always infanticide, pure and simple.

15. Figures from Harris, *Cannibals and Kings.*

16. Hrdy, *Mother Nature,* p. 339. Hrdy herself provides a source for this sentiment. She is by no means a knee-jerk ultra-Darwinian, though her interpretations of infanticide coincide most strongly with received gene-centered evolutionary biological theory.

17. Langer, "Infanticide: A Historical Survey," *History of Childhood Quarterly* 1 (1974): 353–65.

18. Hrdy, *Mother Nature,* pp. 325–27, 338ff.; the quotation is from p. 338.

19. Inability to pay a dowry for one's daughter is an oft-cited reason why people choose not to have daughters or to eliminate (usually through neglect) the ones they do have. For many Indian families these days, it is simply cheaper to have the ultrasound test and abort the baby than to give birth to the girl infant—and medical clinics have used this hard economic fact in their advertising. While prospective economic pain is no doubt part of the reason for female infanticide, I have yet to encounter any examples of selective male abortion/infanticide by families thought to be worried about meeting bride price customs when it comes time for their sons to be married.

20. Harris, *Cannibals and Kings,* pp. 40ff., citing the anthropological literature; the quotation is from p. 50.

21. Singapore is a city-state, having both a mayor presiding over

the core inner city and a prime minister for the entire state. Recent reports on Singapore's fertility issues quote the current prime minister, Goh Chok Tong, while media reports in the 1980s consistently quoted the mayor at that time. Data and recent history of Singapore are drawn from a website article entitled "Singapore Fails," *www.thecore.edu.sg/sea/students/bbonus*, downloaded December 2002.

22. "Celebrating One Hundred Years of Failure to Reproduce on Demand," an Editorial Observer piece in the *New York Times*, April 14, 2002.

23. How ironic, then, that many of the conservative right in the contemporary United States, bent upon excusing, even legalizing, some of the most ruthless elements of corporate behavior (like watering down long-agreed-upon retirement benefits), are staunch opponents of evolution. However dimly perceived and distorted the fundamentals of the evolutionary process are, they remain the ultimate rationale for competitive human economic behavior even among those who would prefer not to acknowledge that life—especially human life—has evolved.

24. Data from *Pocket World in Figures*, 2002. "Crude birth rate" is defined as the number of live births in one year per 1,000 population, while "fertility rate" is the average number of children borne by a woman throughout her reproductive years.

25. Statistics from several Internet sources, including the pro-Palestinian website Hanthala Palestine, in an article entitled "1948 Palestinians: The Statistics." I freely acknowledge that, in the continued tragic heat of the Palestinian/Israeli conflict, all such sources are automatically suspect. Yet the numbers are claimed to have been derived from Israel's Central Bureau of Statistics; in any case, they seem to agree with similar figures given elsewhere.

26. Report issued by the U.S. Census Bureau in October 2001.

27. See Susan M. Essock-Vitale, "The Reproductive Success of Wealthy Americans," *Ethology and Sociobiology* 5 (1984): 45–49.

28. Richard Dawkins, in *The Extended Phenotype* (Oxford and San Francisco: Freeman, 1982), refers to organisms as "vehicles"—the outer packaging necessary for the care, feeding, and aid in dissemination to what's *really* important: those crafty genes.

29. This is the distinction called "r" versus "K" selection in biology: "r"-selected species are those with many offspring and little or no parental investment; "K"-selected species produce fewer offspring,

but generally higher levels of parental investment in the young after birth, hatching, etc.

30. This calls to mind the preflight safety instructions that always tell you to put your oxygen mask on first, and *then* help your kids with theirs, the assumption being that a conked-out adult poses a greater risk for more lives than an unconscious child.

31. A recent CNN story on baby selling in India featured a woman who sold her daughter for twelve English pounds (babies sell for as much as fifty pounds). The woman said she didn't want money at all, but merely a "better life" for her daughter (adoption in the United States was the supposed destination of the child).

32. Ironically, stepping outside of local ecosystems with the invention of agriculture removed limits to population growth; but humans figure ultimately to be limited by the productivity of the global ecosystem—out of the frying pan, into the fire!

## Chapter 11: A Moral to the Story

1. At any rate, so says Richard Dawkins in his essay "A Devil's Chaplain," reprinted in his *A Devil's Chaplain* (Boston: Houghton Mifflin, 2003). Dawkins, pronouncing himself in agreement with George Williams, and with T. H. Huxley (who is quoted as having said, "Let us understand, once and for all, that the ethical progress of society depends, not on imitating the cosmic process, still less in running away from it, but in combating it"). Dawkins goes on to say (p. 10), "As an academic scientist I am a passionate Darwinian, believing that natural selection is, if not the only driving force in evolution, certainly the only known force capable of producing the illusion of purpose which so strikes all who contemplate nature. But at the same time as I support Darwinism as a scientist, I am a passionate anti-Darwinian when it comes to politics and how we should conduct our human affairs." To which I can only say that the perceived nastiness of Darwinism lies in direct proportion to the degree to which one is wedded to the image of the selfish gene.

2. As reported, for example, in Richard J. Herrnstein and Charles Murray, *The Bell Curve: Intelligence and Class Structure in American Life* (New York: Times Books, 1995), whose concluding chapter

expressly discusses the policy implications of the alleged biological differences in intelligence between different races.

3. Dawkins, *A Devil's Chaplain*. The passage in question occurs in his essay "Unfinished Correspondence," p. 221ff.

4. Quotation from Williams, *Plan and Purpose in Nature* (New York: Basic Books, 1996), p. 157. The passage deals with the issue of wrestling with the supposed moral implications of this vision of Darwinism—as discussed earlier in this chapter.

# Index